The Power of User Data: Unlocking Insights and Personalization Across Industries

Jemison

Contents

3   Single-Attribute-Level Problem: Enabling Human Mobility for Socioeconomic Status Estimation   42

4   Multi-Attribute-Level Problem: Multiple Socioeconomic Attributes Estimation based on Home Location   65

Chapter 1

Introduction

1.1 Motivation

User attributes refer to a person's various demographic characteristics, like age, gender, income, education, etc. For academia, user attributes are the basic data sources for many research areas like sociology and economics. For governments, user attributes like income can offer detailed population information for designing and evaluating social policies[16]. Recently, more and more companies have also become interested in leveraging user attributes to promote diverse commercial applications. For example, a user's age, gender and income can help recommender system to understand the user's preferences and provide more personalized services[91, 19, 51, 102].

The traditional methods to collect individual-level attributes for a population are large-scale manual survey, including household interviews, telephone interviews, online questioners, etc. Although traditional methods can get detailed and accurate user-level information, they are highly expensive and time-consuming. And, the time gap between two successive large-scale surveys could be very long, which may be several months or even several years[12].

Fortunately, the burst of available user-generated data provides researchers another way to infer attributes. Nowadays, billions of people all over the world are generating massive data every day, such as online shopping data, online check-in data, smart card mobility data, social media data, etc. For example, according to [67], about 1.9 billion people are purchasing online in 2019; according to [62], more than 10 million passengers travel daily in subway systems in cities like Shanghai. These user activities keep generating data that reflects people's lifestyle and personal habits. And user attributes are related to people's lifestyles and habits. Thus inferring user attributes via mining user-generated data has become an important research area, attracting more and attention from data mining fields.

1.1.1 Open Problems

Compared with manual survey methods, data mining methods are cheaper and quicker. So a variety of methods have been proposed to infer different user attributes from various user-generated data. For example, [12] estimates Rwandans' wealth based on their mobile phone usage data. The experiments show that the distribution of wealth estimated from mobile phone data has a strong correlation with the distribution of wealth measured by the Rwandan government. [72, 73, 58] explore how to estimate people's job types based on their tweets contents. Although some promising discoveries have been made in the field of UAI, there are many open problem requiring further discussion.

Single-Attribute-Level Problem of UAI

The first problem is to **enable new kinds of data sources to infer attributes**. More specifically, in this book, we focus on enabling human mobility data to infer people's socioeconomic status (SES) [86].

Human mobility is an important kind of people's real-world behavior. Many mobility datasets have been gathered and opened for research and commercial usage [63, 71]. Existing SES inference works mainly rely on people's cyberspace data like tweets. These UAI methods cannot estimate attributes if they do not have the cyberspace data (e.g., tweets) of target users. For example, it is much harder for public transit agencies to collect travelers' cyberspace data than mobility data. For users with cyberspace data, mobility data may also help to further improve the performance in attribute inference. Last but not least, the study on mobility-based SES prediction can also help to understand to the underlying relationship between people's mobility patterns and SES. So introducing human mobility data into SES inference is an important open problem for UAI.

Multiple-Attribute-Level Problem of UAI

The second problem is to **improve the accuracy in inferring multiple attributes with limited data sources**. More specifically, in this book, we focus on improving the accuracy of inferring multiple socioeconomic attributes like income and education level when the only raw input data source is the location address of people's homes.

In real-world scenarios, there may be obstacles like privacy law and budget limitations during the raw data collection phase. So it is quite common that UAI researchers only get limited generated data of target users. It is hard for machine learning methods to get accurate estimation when the input effective information is limited. What makes it harder, UAI researchers are often required to estimate as many attributes as possible. Because in this way, the value of raw data collection can be maximized. Different from the first problem, we can still get an estimation of user attribute. Our aim is to improve its accuracy.

Socioeconomic attributes inference is an important problem in social computing. The existing methods can get satisfactory estimations with rich input data sources (e.g., hundreds of tweets). However, there is few discussion about improving the accuracy with limited input data sources (e.g., a single geolocation address of user's home). So improving the accuracy of predicting multiple socioeconomic attributes from home location is an important open problem for UAI.

Multiple-Task-Level Problem including UAI and UAE

The third problem is to **improve the performance User-Attribute-Enhanced (UAE) tasks by UAI**. The first two open problems mainly consider the performance of UAI itself. The performance of UAI is very important to User-Attribute-Based (UAB) tasks. However, it is not the main concern for UAE tasks.

For UAB, user attributes are the basic data. UAB tasks cannot be carried out if the attributes cannot be estimated (like the first open problem) or the accuracy of estimated attributes is too low (like the second open problem). For example, content-based recommendation (CB) is a typical UAB task. It recommends items based on users' attributes. So if the missing attributes cannot be estimated or the accuracy of the estimated attributes level is too low, we cannot use CB to recommend items to users.

The other kind of tasks are the UAE tasks. User attributes are just the auxiliary/supportive input data for UAE. UAE tasks can get a result without any user attributes. But if there are complete attributes, the performance of UAE tasks can be further improved. Collaborative filtering recommendation (CF) is a typical UAE task. CF recommends items based on users' behavior history. If the behavior data is not enough, user attributes can also help to improve the performance of CF.

Until now, many UAE tasks (e.g., all CF methods), do not consider UAI at all when there are missing attributes. If one kind of attribute is incomplete, they usually ignore the kind of attribute or use zeros as substitutes for missing values. This is because UAE methods often face the problem that there are many kinds of missing attributes. As we can see in the discussion of the first two problems, to design a high-accuracy UAI method for various missing attributes require a lot of work.

Besides, there is no grantee that the contribution of those work to the final performance of UAE will be s ignificant. So the tr aditional idea to improve the performance of UAI is not suitable for UAE. Until now, UAI is often overlooked by UAE even with missing attributes, which leads to the sub-optimal result of UAE.

Recently, more and more UAE tasks emerge, such as recommendation and advertising. For some UAE tasks (e.g., recommendation), even a small improvement can be important for researchers or companies. So how to improve the performance of UAE with UAI become more and more important. In this book, we focus on leveraging UAI to help improve the performance of CF-based recommender systems.

1.2 Research Goals

In this book, we aim to address the three UAI problems dis-cussed in 1.1. The corresponding research goals include:

G1 Our first research goal is to design a deep-learning-based method that can predict people's SES based on one kind of human mobility data – smart card data (SCD).

G2 The second goal is to develop a multi-task learning method which can improve the accuracy in predicting people's personal income, family income, educational level and job types from a limited input data sources: home location.

G3 The third research goal for us is seeking to design a unified model to improve the performance of CF tasks with auxiliary UAI tasks.

We can see that the first research goal is a single-attribute-level work that focus on the accuracy of one specific attribute: SES. The second is a multi-attributes-level work which tries to improve the accuracy of various attributes. The third one further extends from multi-attributes-level prediction into multi-tasks-level work which considers UAE as well as multi-attribute-level prediction. These three research goals show a way to extend UAI from a single-output task which mainly serves for UAB tasks, to multi-output tasks which serves both UAB and UAE tasks.

Section 1.3 and Section 1.4 will explain the challenges in reaching the research goals and our main contributions respectively.

1.3 Research Challenges

1.3.1 Single-Attribute-Level Problem: Enabling Human Mobility for Socioeconomic Status Estimation

In this section, we discuss the challenges in achieving the first research goal. Before that, we need to give a short introduction about the background knowledge of SES inference and smart card mobility data.

Background

SES is a widely studied concept in the field of social sciences[86]. Unlike simple attributes like gender or age, it is an economically and sociologically combined overall measure of an individual or family. SES can be calculated based on one or several basic indicators like people's income level, education level and job types. It describes one's economic and social position in relation to others and is typically divided into three levels (high, middle, and low)[86, 16, 84]. An individual with a higher SES means he/she earns more, has a better job or higher education than those with a lower SES.

SES can provide measurable inputs for related studies like social stratification, social welfare and business planning[87, 12]. Nowadays SES is not limited to social sciences. It also becomes important to governments during designing social policies. And SES also begins to be used in many commercial applications, like personalized recommendation, customized marketing and precise advertisement [91, 19, 51, 102].

Companies and governments often need to get the SES information of a large population. However, the cost of manually collecting SES information of a population is unbearable to most companies and even governments in some developing countries[12]. Thus recently, UAI researchers have become more and more interested in SES inference. Until now, some efforts have been made to estimate individual-level SES using cyberspace data sources like online social media [72, 73, 58].

These data-based methods can lower the cost in getting individual-level SES of people, if their social media data could be found by researchers. However, sometimes researchers cannot find social media data for targeted users. So we need to keep introducing new kinds of data sources for SES inference to cover as many users as possible.

Among potential new data sources, human mobility data can be of great help. Because mobility patterns can be used to describe one's lifestyles. And data-based SES inference methods are actually based on the assumption that different SES levels of people have different lifestyles.

Specific Challenges

In this book, we choose the Smart Card Data (SCD) of Shanghai city as a case study of the human mobility data source. SCD is generated by smart card automated fare collection systems. The automated fare collection systems are now widely used by public transit agencies around the world [9, 64]. The dataset is opened by the Shanghai government and includes a great amount of individual-level, time-stamped and geo-tagged trip data of Shanghai citizens.

Although many previous works have studied SCD or SES inference, the discussion about estimating SES based on SCD is quite limited for following challenges:

- The first challenge is that it is hard to collect ground-truth SES data for large-scale SCD users. As far as we know, there are no open datasets that consist of people's mobility (e.g., SCD) and their SES data simultaneously. UAI is a data-based supervised learning problem, which needs ground-data SES label for training machine learning models. So we first need to get the SES labels for millions of users in Shanghai SCD.

- The second challenge is to design effective SCD-based features that may reflect people's SES levels. This is also the basic problem for any UAI work which tries to enable new data sources. There are some cellphone-data-based methods [87, 105, 12] which can predict group-level SES. They discussed some general statistical mobility features, like the average daily moving distance. However, these works mainly rely on cellphone features like the numbers of calls and telephone fares. The general statistical mobility features are just supportive information. So the mobility features are not effective enough for organizations (e.g., public transit agencies) which only have human mobility data. Besides general statistical mobility features, we need to design new SCD-based features that can effectively capture the dynamic urban lifestyle of subway users in Shanghai.

- The third problem is closely related to the designed features. Existing SES inference methods mainly rely on standard classical machine learning methods like support vector machine (SVM), Gradient Boost Decision Tree (GBDT) or Multi-Layer Perceptron (MLP). They are good at processing statistical mobility features. However, they may be not suitable for dynamic sequential input features. So we need to design a model that can utilize both statistical and dynamic sequential mobility features to improve the accuracy of SES inference.

1.3.2 Multi-Attribute-Level Problem: Multiple Socioeconomic Attributes Estimation based on Home Location

In this section, we discuss the challenges in achieving the second research goal. Before that, we will deliver an short introduction about background of socioeconomic attributes inference.

Background

In this book, socioeconomic attributes mainly refer to people's income level, education level and occupation types. Inferring individual-level socioeconomic Attribute (SEA) is an important problem for social computing [4]. Like SES, these attributes also play an important role in studies like social stratification and social welfare. And they are also the basic factors to calculate people's Socioeconomic Status (SES) [16, 84]. Compared with SES, these indicators are much easier to be understood. They have already been widely used by people who are not experts in sociology. In particular, online service providers pay special attention to SEAs if they want to offer personalized services in recommendation and advertisement [91, 19, 51, 102].

Until now, there have been a lot of works in inferring SEA for a large population. [87, 12, 5, 105, 72, 73, 58]. For example, [72, 73, 58] explore how to estimate people's income or occupation based on the language patterns, topics or even emotions in tweet content. [87, 12, 5, 105] focus on predicting peoples family income from their mobile phone usage habits. [96, 68] estimates people's income and education level based on how people purchase items in offline retailers.

These methods could get accurate SES levels from the rich information contained in data sources like tweets or cellphone data. However, these methods did not discuss the problem that the data sources only contain limited information. Actually, in real-world scenarios, it is quite common that researchers can only rely on limited input information. For example, a large part of users has few or no tweets content or cellphone records at all. Sometimes, limited by budget and time, researchers or companies could only get a kind of user-generated data that contains very few useful information. Here we choose the home location as a case study of limited input data sources.

Specific Challenges

There are several challenges of investigating the relationship between people's SEAs and home location:

- The first challenge is similar to SES inference, no open datasets are containing both personal SEAs and home location. We need to collect our own datasets before designing any data-mining-based methods.

- The second challenge is that home location itself only contains limited information. The accuracy of SEA inference would be low only based on the latitude and longitude of people's homes. What makes it worse, income, occupation or education levels are all complex attributes that are hard to predict even with rich human behavior data like in [104, 12]. We need to enrich the home location with more SEA-related knowledge by feature design and data mining.

- The third challenge is to design a machine learning method which can generate new interacted features based on the basic input feature. And the method should also be able to increase the accuracy in estimating one attribute by the other attributes.

    Chapter 1 Introduction

1.3.3 Multi-Task-Level Problem: Improving User-Attribute-Enhanced tasks by Attribute Inference

In this section, we discuss the challenges of improving the performance of UAE tasks with UAI. In this book, we pick a typical UAE task, CF recommender system as a case study. First we need to introduce the background of UAE and recommender systems.

Background

Nowadays, online users often find that there are too many kinds of books, movies or songs to choose. The recommender system is a crucial tool to help users to find what items they may prefer to interact or buy[78]. Until now, collaborative filtering (CF) is one of the mainstream recommender systems [79, 27, 81]. CF bases on an assumption that a user would tend to like items that are liked by the other "similar" users. CF measures the similarity of users based on their interaction histories with different items.

Recently, CF methods begin to leverage an emerging machine learning method, Graph Convolutional Network (GCN) to [11, 106, 98, 45], to improve the performance. For instance, GC-MC [11] applies GCN on user-item graph to exploit the direct connections between users and items. NGCF [98] improves the recommendation performance by modeling high-order connectivity on a user-item graph. And LightGCN achieves state-of-the-art performance by simplifying feature transformation and nonlinear activation in GCN layers [45]. Most of these CF methods do not consider user attributes.

Sometimes CF methods may encounter *interaction sparsity problem*. This is because many users may only interact with a very small proportion of items. The few interactions of these users are insufficient for CF to learn their accurate preference for items. To alleviate the problem, researchers tried to use various attributes of the user (e.g., gender, age, location) and item (e.g., category, genres, brands) to improve the original CF methods [82, 57]. For CF methods also leveraging attributes, we refer them as attribute-enhanced CF methods. These CF methods are typical UAE tasks. They can still recommend items without any attributes. And if they can get the attributes of the users who have few interaction histories, the performance can be further improved.

In real-world recommending scenarios, user/item attributes are often incomplete. For instance, many users are reluctant to provide age or location information due to privacy concerns. Until now, there has been limited discussion of leveraging UAI methods to estimate these missing attributes for CF methods. This is because there may be usually many missing attributes in recommender systems. From the first two challenges of our book, we can see that to design an accurate UAI method for various missing attributes requires a lot of work, such as data collection, data mining, and new feature/model design. However, the final contribution of these UAI-related works to the recommender results is not clear. Especially, when the missing ratio of one attribute is too high, it is very hard to design an attribute inference method even for UAI experts.

Until now, CF researchers simply use zeros, average values, or special tags as substitutes for missing values, without specially designing UAI methods. These simple substitutes can make attribute-enhanced CF methods easily adaptive to incomplete attribute features, though their performance will be affected if the missing rate is high.

Specific Challenges

As far as we know, there is no discussion about unifying UAI into CF methods to improve the recommending performance. Because there are several challenges to reach this research goal:

- The first challenge is that we need to quantify the effect of ignoring UAI methods to the recommending performance. This is essential because it can show the value of combining UAI and CF tasks to both UAI and CF researchers. Without this quantifying experiments, these two groups of researchers may continue to focus on their area and are not interested in interaction.

- The second challenge is to design a new framework to lower the cost of trying UAI methods for CF tasks. For different CF datasets and methods, there are various kinds of missing attributes. Some missing attributes can be important to the recommending performance and some are not. We cannot afford to try to design accurate and sophisticated UAI methods to predict all attributes at first and then find out some attributes that are not useful to the recommendation at all. The framework should be suitable for various amounts and missing ratios of attributes, and can quickly find out which attributes are really needed. And the framework can predict the attributes based on existing interaction data without requiring extra data collection or data mining works.

- The third challenge is to control the influence of UAI to CF if the accuracy of UAI methods is not high. Actually there are many cases that the accuracy of UAI would be low. For exmaple, the missing ratio of the attribute may be too high (more than 90%). It is hard to get accurate estimation if most labels are missing for any machine learning problems. Second, some kinds of attributes are too hard to be estimated if we only rely on users' interaction history. Lastly, even we know how to increase the accuracy of UAI, soemtimes we have to give up to control the cost. If the accuracy of estimated attributes is too low, UAI will misleads the CF results instead of improving CF. In this case, the recommending performance maybe even worse than simply ignoring the attributes. So we need to design a mechanism to dynamic control the influence of UAI to CF.

1.4 Summary of Contributions

This section describes the main contributions during realizing the three research goals described in section 1.3.

1.4.1 Single-Attribute-Level Problem: Enabling Human Mobility for Socioeconomic Status Estimation

To tackle the challenges described in chpater 1.3.1, we propose an SCD to SES (*S2S*) method to infer people's SES from their SCD mobility data. To the best of our knowledge, this is the first attempt to estimate individual-level SES from SCD data. Our main contribution is summarized as follows.

- **Ground Truth Construction**. The dataset we studied in this book (nearly 8 million smart card IDs) is totally anonymous. We cannot manually relate any user ID to their actual SES levels. First, We carry out a survey in Shanghai and find out there is a strong correlation between the housing price level and income levels in Shanghai. Then we analyze the mobility data of all users and observe that the main part of the smart card dataset comes from a part of users who often take subways. We analyze the mobility patterns for these frequent users and identify their working and home locations. Then we mined housing prices from multiple commercial real estate websites. In the end, the housing price level of home location is chosen as the proxy ground truth for these frequent users.

- **Sequential Feature Design**. We observed that: 1) people of different SES may visit different places and have different commute schedules; 2) people show in the different functional areas may have different social attributes. After dividing all subway stations into 3 kinds of function areas, we designed a new sequential feature that describe when and which function areas people travel every day.

- **Model Design and Experiments**. We propose a deep neural network (DNN)-based learning model (S2S), which combines the mobility information from both sequential features and general statistical features. The experiments on the large-scale smart card dataset in Shanghai City demonstrate that: 1) the proposed method can use the human mobility data to estimate SES level; 2) S2S significantly outperforms widely used baselines like Xgboost; 3) the sequential features and corresponding component of S2S model represent more salient nature of an individual's behavior in socioeconomic context than traditional general statistical features.

1.4.2 Multi-Attribute-Level Problem: Multiple Socioeconomic Attributes Estimation based on Home Location

To tackle these challenges described in chapter 1.3.2, we propose a home to SEA (*H2SEA*) method to infer multiple individual-level socioeconomic attributes from people home location. To the best of our knowledge, this is the first work focusing on SEA inference through the home location. The main contributions are summarized as follows:

- **Design and mine data for Home-based Features**. We extend people's home locations with more knowledge from various aspects such as area-level economic statistics, housing price, point of interest (POI), and administrative division. Multiple SEA-related features are designed according to this knowledge. The source data of these features are mined from multiple commercial real-estate websites, official statistic bureau websites, online maps, etc.

- **Feature Interaction, Multi-task Model**. We propose a factorization-machine-based multi-task learning method with an attention mechanism, to learn a shared representation from input features as well as attribute-specific representations for different SEA predication tasks. The multi-task method can additionally leverage the potential relationship between income, education and occupation. Comparing with existing multi-task learning methods for attribute inference, the proposed model further improves the performance with limited features by modeling the second-order feature interactions with factorization machine (FM).

- **Dataset Construction and Experiment**. We carry out a large-scale survey to collect people's personal income level, family income level, occupation types and education level in China. In the end, we collect a dataset that includes 9 provinces and 85 cities in China. The experiments on this dataset demonstrate that 1) home location can improve the performance of predicting people's SEAs; 2) the proposed method outperforms compared methods on all SEA prediction tasks in terms of multiple metrics such as AUC and F1-measure.

- **SEA-Home Relationship Analysis**. By further analyzing the relationship between SEAs and home location, we made several interesting observations: 1) home location is more helpful in predicting personal income than family income; 2) the most important features in most SEA predictions are county-level average income and POI distribution instead of housing price. We find out that these are caused by a weaker relationship between housing price and income level in China.

1.4.3 Multi-Task-Level Problem: Improving User-Attribute-Enhanced tasks by Attribute Inference

To tackle these challenges described in chapter 1.3.3, we propose AEGCN, an end-to-end multi-task GCN-based CF method, which improves recommending performance with incomplete attributes by auxiliary user/item profiling tasks. To the best of our knowledge, this is the first framework that combine UAI into CF (UAE) tasks to improve the recommending performance. The main contributions are summarized as follows:

- **Show the value of UAI for CF**. We highlight the problem of the missing attributes by quantifying the negative impact of the missing attributes on recommending performance through empirical studies. We choose three real-world large-scale recommending datasets and compare the performance between complete and incomplete attributes. For the incomplete attributes, we simply use unknown tags as substitutes instead of using UAI methods to the missing values. The comparison results clearly show that the recommending performance is more and more seriously affected by increasing missing rates.

- **Framework Design**. We propose AEGCN, an end-to-end multi-task GCN-based CF method, which improves recommending performance with incomplete attributes by auxiliary user/item profiling tasks. User/item attributes can be predicted based on user-item interactions, which is also the source data for recommending. From the perspective of GCN, the two tasks are both graph node representation learning tasks by modeling node interactions. The estimation from user/item profiling task is usually more accurate than simple substitutes. Thus it can alleviate the problem of the missing attributes for recommendation by taking user/item profiling as an auxiliary task.

- **Experiments on Large-scale Real-world Datasets**. We conduct extensive experiments on three real-world datasets. which demonstrates the effectiveness of AEGCN in alleviating the problem of the missing attributes. When the missing rate increases, AEGCN consistently outperforms state-of-the-art Collaborative filtering (CF) models without attributes. Compared with other attribute-enhanced CF models, AEGCN achieves comparable performance when the attributes are complete, and significant improvements when the missing rate increases.

1.5 Book Outline

This book contains the content appearing in the following published and submitted papers.

- Shichang Ding, Hong Huang, and Xiaoming Fu. Estimating Socioeconomic Status via Temporal-Spatial Mobility Analysis-A Case Study of Smart Card Data. International Conference on Computer, Communication and Networks (ICCCN 2019).

- Shichang Ding, Xin Gao, Yufan Dong and Xiaoming Fu. "Estimating Multiple Socioeconomic Attributes via Home location – A Case Study in China." Under submission.

- Shichang Ding, Xiangnan He, and Xiaoming Fu. AEGCN: Attribute-Enhanced Graph Convolutional Network for Recommendation with Missing Attributes. Under submission.

Shichang Ding's contributions to each papers are as follows.

- For the first paper, Shichang Ding proposed the idea of this work, designed the features, collected the data for features and labels, designed the algorithm (S2S), carried out the experiments, and wrote the original draft. Xiaoming Fu provided the smart card dataset. He and Hong Huang both revised the draft. Xiaoming Fu also acquired all the funding needed for this project.

- For the second paper, Shichang Ding conceived the idea, designed the features and the algorithm (H2S). He also performed the experiments and wrote the original draft. Xin Gao provided the dataset consisting of people's socioeconomic attributes. Shichang Ding and Yufan Dong collected the data for features. Xiaoming Fu revised the draft, supervised the project and acquired the funding for this work.

- For the third paper, Shichang Ding and Xiangnan He proposed the idea. Shichang Ding designed the algorithm (AEGCN), carried out the experiments, and wrote the original draft. Xiangnan He provided the computing resources for the experiments, funded Shichang Ding during his visit in university of science & technology of china. Xiangnan He and Xiaoming Fu both revised the draft. Xiaoming Fu supervised the project.

The outline of this book are as follows:

- In Chapter 1.1, we first briefly introduce the background and three open problems of user attribute inference (UAI). In Chapter 1.2, we describe our research goals to overcome the open problems. Then in Chapter 1.3, we present the main challenges to fulfill each research goals. In Chapter 1.4, we concludes the main contributions corresponding to each open problem. In chapter 1.5 we outline the organization structure of this book.

- In Chapter 2 we reviews the existing studies related to three problems discussed in this book. For single-attribute-level problem, chapter 2.1 presents the related works about SES inference in chapter. For multi-attribute-level problem, chapter 2.2 presents the related works of SEA inference. For multi-task-level tasks including both UAE and UAI, chapter 2.3 presents the existing studies for CF recommendation.

- In Chapter 3, we take smart-card-data-based SES inference as a case study of single-attribute-level problem. We present a deep neural network (DNN)-based learning approach (S2S) to infer personal SES from his/her smart card data. The method considers both temporal-sequential features and general statistical features of human mobility. More specifically, In Section 3.1 we propose the motivation, challenges and contributions of S2S. Section 3.2 introduces the smart card datasets. Section 3.3 discusses the temporal-sequential features and general statistical mobility features. The detail of S2S model is discussed in Section 3.4. Experimental results on Shanghai smart card datasets are presented in Section 3.5 . The work is concluded in Section 3.6.

- In Chapter 4, we take Home-based multiple SEA prediction as a case study of multi-attribute-level problem. In this Chapter, we propose H2SEA, a deep learning method which can predict a person's multiple socioeconomic attributes from home location. To be more specific, Section 4.1 introduces the motivation, challenges and contributions of this work. Section 4.2 introduces the ground-truth dataset collected in China. Section 4.3 discusses how to design and mine data for Home-based SEA-related features. The H2SEA model is proposed in Section 4.4. Experimental results are presented in Section 4.5. Section 4.6 further analyzes the relationship between housing price and income in China. The conclusion of this chapter is in Section 4.7.

- In Chapter 5, we take CF recommender system as a case study of multi-task problem. In this chapter, we proposes AEGCN, an end-to-end multi-task GCN-based CF method, which improves recommending performance with incomplete attributes by auxiliary user/item profiling tasks. The motivation, challenges and contributions of the work in this chapter are firstly introduced in Section 5.1. Then Section 5.2 we give an detailed description of AEGCN model. The efficiency of AEGCN model is evaluated in Section 5.3. Finally, in Section 5.4, we conclude the work in this Chapter.

- In Chapter 6, we summarize the three works in this book and discuss the possible future research work.

Chapter 2

Literature review

In this chapter, we review the existing studies on User attribute Inference. For single-attribute-prediction level, we present the corresponding works about SES inference in chapter 2.1. For multiple-attribute-prediction level, we introduce the related works of SEA inference in chapter 2.2. For multi-task-level tasks including both UAE and UAI, we describe the existing studies for CF recommendation in chapter 2.3.

Contents

2.1 Socioeconomic Status Inference

SES is a widely studied concept in the field of social sciences, especially in health and education analysis [16]. In recent years, companies and researchers pay increasing attention to SES estimation because of its potential in numerous high-value applications like personalized recommendation and online banking. Though there has been a great improvement in estimating other demographic attributes like age, ethnicity, and gender [112, 7], SES estimation still needs more effort. One of the main obstacles is that SES ground truth data (covering a large group of people) is much harder to get than attributes like age and gender. Normally users are more reluctant to disclose their education, occupation, and income information. The organizations, which have such data, also seldom open it to the public for privacy reasons. Recently, researchers begin to use indirect SES indicators from some big data sources. These data sources may cover millions of people, recording different aspects of their lifestyles.

2.1.1 SES Estimation based on Social Media

Social media is an important cyberspace user-generated data source that researchers pay a lot of attention to. Preotiuc-Pietro et al. present the first large-scale systematic study on inferring individual-level occupational class, which is quite similar to SES, from user-generated data on social media[72]. In this work, they mainly focus on users' language use on social media. They collect 5,191 English users who mentioned their occupation in the user description field. And these users all at least have more than 200 tweets. Then they design user-level textual features based on users' aggregated set of tweets, through singular value decomposition (SVD) word embedding, normalized point-wise mutual information(NPMI) clusters, neural embedding, and neural clusters. In the end, they used a non-linear Gaussian Process (GP) framework to estimate users' occupation class. The experiment results highlight that a user's occupation influences his/her language use pattern.

Lampos et al. present one of the first methods for inferring the individual-level socioeconomic status of social media users[58]. They collect 1,342 English users' profiles from Twitter. The users are selected based on whether they report the occupation type in the profiles. Researchers then collect tweets of these users from February 2014 to March 2015. Researchers calculate users' SES based on occupation types. Compared with [72], they add other non-textual features like the total number of tweets and the number of accounts followed, etc. These features characterize users' platform-based behavior and their importance on the platform. In the end, researchers also use GP to predict people's SES from the user-level social media features.

Huang et al. want to analyze the relationship between SES and people's activity patterns extracted from Twitter[50]. Researchers collect 7,660 users who live in Washington, DC, and have more than 40 geo-tagged tweets. Then these users' home and working areas can be inferred based on the geographical and temporal information of these geo-tagged tweets. Then researchers analyze users' activity patterns, which mainly include the number of activity zones, distance between home and activity zones, standard deviational ellipse, etc. From these activity patterns, they find out that while SES is highly important, the urban spatial structure also plays a critical role in affecting the activity patterns of users in different communities.

Abitbol [1] proposed a method to infer the SES of Twitter users, combining information from numerous sources, including Twitter, census data, LinkedIn, and Google Maps. First, they collect more than 90 million tweets, posted by 1.3 Million French users over one year. Then they find the home location of users based on the geo-tagged tweets. In this way, they map users to census blocks. The median income of each census block is published by the National Institute of Statistics and Economic Studies (INSEE) of France. The median income of a census block is used as an approximation of the income level of the Twitter users live in it. Users' occupation data can be found if they provide their LinkedIn account in their tweets or profiles. Researchers also estimate the socioeconomic features of users' living area by users' street views from Google Earth. They invite experts to annotate the level of users' living area by watching the street views. A user's SES level is the combination of census income data, occupation data, and housing price data. The features are similar to previous works like [58, 50], including users' profiles and textual features extracted from tweets. In the end, researchers use three classical machine-learning methods (AdaBoost, Random Forest, and XGBoost) to predict users' SES levels.

2.1.2 SES Estimation based on Cell Phone Data

Another important user-generated data type is mobile phone data. However, most of the existing studies only focus on group-level SES inference (at least until the acceptance of our work [25] in 2019). Soto et al. explore how to use information derived from the aggregated use of cell phone records to identify the socioeconomic levels of a population [87]. More specifically, their work can get a socioeconomic level to the area of coverage of each base transceiver station (BTS) tower. In the city, a BTS can cover about 1 square kilometer of areas. Researchers only study the users who frequently call otherwise the information of users is not enough for analysis. They design various features of users' calling behaviors to distinguish each BTS tower. The features include the aggregated calling behavior of one BTS area, like the total number of calls or short messages. The SES of a BTS area is calculated based on the published house-hold income, occupation by governments. In the end, Soto et al. use standard classical machine learning methods such as Support Vector Machine (SVM) and random forests to predict the SES of each BTS area. Though this method is one of the first to predict (group-level) SES from cell phone data, it cannot estimate the individual-level SES of each person. It is not a UAI task.

Based on the same datasets, Frias-Martinez et al. then explore the relationship between various features of cell phone usage (including mobile phone consumption, social information, and mobility patterns) and socioeconomic indicators (including income and education) [30]. They find that a person's SES is moderately or strongly correlated with his/her average calling physical distance, cell phone-related cost, exchange frequency of communications, and frequently-traveled geographic location.

Blumenstock et al. propose a method to estimate a finer-grained group-level SES (i.e., household-level)for Rwandans based on cell phone data[12]. The researchers first design a composite wealth index for Rwandans based on whether they have refrigerator, electricity, television, and other belongings. The data is collected through a telephone survey. Then they extract features from the mobile phone data. In the end, they use a standard classical machine learning method to estimate people's wealth indexes from these features. The experiments show that the distribution of wealth estimated from mobile phone data has a strong correlation with the distribution of actual wealth measured by the Rwandan government. This work considers multiple factors of phone usage including communication, the structure of and contact network. The mobility pattern is discussed as a supportive feature. Different from them, we mainly rely on mobility features and use a different kind of data source (SCD).

Almaatouq et al. propose a method to estimate the district-level unemployment rate from people's mobile communication patterns[5]. The average spatial resolution of the district is less than 2.7 km. The ground truth data comes from an unemployment benefit program. They also find that aggregated calling activity, communication networks are strongly correlated with unemployment.

Yang et al. analyses the relationship between multiple mobility features and SES based on mobile phone datasets of two cities: Singapore and Boston[105]. In Singapore, they take the housing price of living areas as SES. In Boston, they use the census tracts as SES. They find that the relationship between mobility and SES could vary among cities, and such a relationship is quite complicated. It may be influenced by several different factors like spatial arrangement of housing, employment opportunities, and human activities. For example, phone user groups that are generally richer tend to travel shorter in Singapore but longer in Boston. Our work in the 3 is different from [105] in the following ways: 1) we examine the extent to which SES can be estimated from SCD, while they try to figure out the relationships between SES and mobile phone mobility data; 2) we mainly focus on SCD instead of mobile phone.

2.1.3 Relationship Study between SES and Smart card Data

In recent years, automated fare collection (AFC) systems have become more and more widely used in cities all around the world[66]. The original aim of deploying AFC systems is to make the charging process quicker and cheaper without manual interference. However, researchers realize that the massive and continuous smart card data recorded every day can benefit many fields. For example, smart card data can be used to understand the demand pattern of public transport. The knowledge is of great help to plan new public transportation system [66]. Smart card data can also be utilized to investigate passengers' travel patterns [110]. However, the work about the relationship between SES and smart card data is quite limited.

Langlois et al.[37] investigate the multi-week activity patterns of 33,026 public transport users in London based on their smart card data. Researchers first represent each passenger as an ordered sequence of activities over several weeks. From the sequence, they can capture information relating to travelers' temporal patterns of journeys. Then researchers cluster users according to each user's long-term activity sequences using k-means algorithms. In this way, they find 11 clusters of London public transit travelers. The long-term mobility characteristics of each cluster are quite different. For example, different from other clusters, users in the first four clusters are more possible to move between the primary and secondary locations during the weekday. Then researchers survey a small part of users (1,973) about their demographic attributes and then analyze the demographic attributes of each cluster. They find that the average incomes of some clusters are higher than the others. This work indicates that income may be related to people's smart card mobility data.

Mohamed et al. introduce an approach to cluster passengers living in Rennes (France) based on their temporal habits[64]. They study how fare type proportions are distributed in different clusters. The Rennes SCD dataset includes fare types like Young subscribers, Regular subscribers, Elderly subscribers, etc. They find out there are some mobility differences between different fare type categories. For example, the clusters mainly consisting of students who tend to get back home early on Wednesday since course hours on Wednesdays end early in France, while other clusters do not have this pattern. This also indicates SCD records may be related to users' age and occupation. These works show there is some possible relationship between SCD-based mobility and SES. In section 3, we aim to explore whether and how SCD can be used to estimate SES.

2.2 Socioeconomic Attributes Inference

In chapter 4, we mainly investigate whether people's home location can be used to infer multiple personal SEAs. Our topic mainly relates to two domains: socioeconomic attributes prediction and multi-task learning.

2.2.1 Personal Socioeconomic Attributes Prediction

Personal SEA inference is a proxy method to collect economic or social statistics in some developing countries [13]. The estimated personal

SEA can also be used to improve personal recommendations and precise marketing. Given its importance, a great number of approaches have been proposed to estimate income level, occupation, and education. As far as we checked, most of them try to predict SEAs from people's cyberspace behavior data, like mobile phone calls [12] and Twitter contents [73].

Taking personal income prediction as an example, the two most widely studied data source types are from online social networks (OSN) and mobile phone (mainly include call detail records and usage data). As shown in Table 2.1, quite a few studies are focusing on OSN-based personal income prediction. Note that we also include part of papers that predict personal Socioeconomic Status (SES). Because SES can be seen as a special version of SEA.

SEA Inference based on Social media Data

Famous OSN platforms like Twitter and Facebook develop fast in recent years. Many important works show that people's SEAs can be predicted by analyzing their tweets, social links or profiles recorded by OSN [1, 73, 72, 58, 92, 93, 4, 40, 94, 95].

Preotiuc-Pietro et al. present the first large-scale study to predict the individual-level income from people's generated social media data [73]. They collect 5,191 Twitter users living in the UK, covering 55 kinds of occupation types. The mean yearly income of each occupation can be found in the Annual Survey of Hours and Earnings [6] published by the British Government. Then researchers design a series of features based on users' profile data and tweet contents, such as perceived psycho-demographics, emotions, and sentiment. In the end, researchers apply Gaussian Process (GP) to predict users' income. The predicted income reach a correlation of 0.633 with actual user income, showing that tweets can be used to predict income. They also analyze how different features relate to the users' income. They find that the percentage of words related to fear or joy, the proportion of retweets, and the topics of tweets are the most important features. For example, higher-income Twitter users are likely to express more fear and anger, whereas lower-income users express more opinions with emotions.

Volkova et al. investigate how to predict Twitter users' income and education level in a series of works[95, 93]. In [95], researchers require workers on Amazon Mechanical Turk to manually check 5,000 Twitter users' online content and profiles. All of these users have posted at least 200 tweets. They need to guess 1) whether a user's yearly income is above 35,000 dollars; 2) whether a user has a college degree. Then they extract textual features from users' tweets. Finally, they leverage a log-linear model to predict these users' income and education levels.

In [93], Volkova et al. improve their method in their last work [95] on a larger dataset. They collect the tweets of 123,513 users from the USA and Canada. They use the model trained in [95] to predict the income and education level of users. The predicted income and education levels are leveraged as estimated labels. Then they extract features that characterize the emotional contrast between users and their neighbor users. Finally, they find both income and education can be predicted based on the emotions expressed by that user and the user's social environment.

Recently, Matz et al. propose a method to predict the income level of Facebook users [61]. Researchers carry out a paid online survey to collect the income information of US Facebook users. Researchers select 2,623 participants who have more than 10 Likes or 500 words in their status updates. Two kinds of data are used for feature extraction: users' Likes on Facebook and the content of Status Updates. A widely used dimensionality reduction method, Singular value decomposition (SVD), is applied to the initial features. And in the end, researchers utilize a commonly-used machine learning algorithm, the ridge regression model, to predict the logged income of Facebook users.

Table 2.1: Related Works of Personal Socioeconomic Attributes Prediction

Work	Source Data	Predicted Attributes
[1]	tweets	SES
[73]	tweets	income
[58]	tweets	income
[93]	tweets	education, income
[4]	tweets	occupation, income
[40]	tweets	income
[94]	tweets	education, income
[95]	tweets	education, income
[14]	tweets	family income
[61]	Facebook Likes	income
[13]	mobile phone metadata	personal income
[87]	mobile phone records	SES
[29]	mobile phone call detail records	income
[12]	mobile phone metadata	income
[90]	mobile phone metadata	income
[8]	cookie	income, education level
[68]	retail transaction records	income,education level
[96]	retail transaction records	income, education level
[25]	smart card transportation records	SES
[74]	WiFi log	education, income

SEA Inference based on Cellphone Data

Another important user-generated data type is mobile phone data. Many existing works try to predict people's income levels based on multiple cellphone-related data like communication, the structure of the contact network, users' mobility pattern, etc.

[87] shows that cell phone calling behavior, social network, and mobility data can be used to identify the wealth level of a population living in a community. The ground truth data is provided by a National Statistical Institute, which considers 134 indicators including the level of studies of the number of cell phones, computers, combined income, occupation of the members of the household, etc.

In [90], researchers propose a method to distinguish whether a person's household is poor or not based on various kinds of cellphone-related data. They first conduct a large-scale country-wide survey in a low human development index country. After the survey, they get more than 80 thousand people's income data and their 3-month raw cell phone data. Then they design 150 features covering basic phone usage data (e.g, calling duration), Top-up transactions (e.g., recharge amount per transaction), social networks, handset type (e.g., the brand of phone), revenue (e.g., the charge of the Internet) and advanced phone usage (e.g., Internet volume). Lastly, researchers use a standard multi-layer feedforward method to predict people's income levels.

In [13], Blumenstock et al. estimate Rwandans and Afghans' family income by extracting features from mobile phone communication extracted and mobility patterns. Researchers find out a model based on the data collected in one country cannot be directly used in another country.

Besides cellphone and social media data, researchers also begin to pay attention to predict SEA based on other kinds of user-generated data like retail transaction records [96, 68]. For example, in [96], Wang et al. present the first methodology to predict users' income and education levels based on in the retail scenario. They collect a dataset from a large retailer in China. The dataset contains more than 49 million transactions between 1.2 million users and 220 thousand kinds of items. Users are represented based on their purchase history. In the end, researchers feed the representation of all users to a log-bilinear model to predict users' income and education levels simultaneously. Different from these works, in section 4 we discuss how to predict multiple sensitive SEAs including income, education, and occupation solely based on people's home location.

2.2.2 Multi-Task Learning for Multi-SEA Inference

Multi-task learning (MTL) is a learning paradigm in machine learning. The main purpose of MTL is to take the advantage of useful information shared in multiple tasks to improve the generalization performance of all the tasks [109]. All of these learning tasks are assumed to be related to each other. Considering the cost of data collection, researchers may need to predict multiple users' attributes from one dataset. Therefore some efforts have been put in studying how to apply multi-task learning in user attribute inference [96, 54].

One of the first multi-task model proposed for socioeconomic attribute inference is Structured Neural Embedding (SNE) [96]. SNE uses a simple dense layer to generate initial embedding vectors for all input features. Average pooling is conducted on these vectors and then fed into a linear prediction layer for each SEA estimation task. Different from the conventional multi-task method, SNE ignores the correlation between attributes. Because they think the correlation is hard to model without explicit knowledge of relationships among tasks. Instead of summation of each task, they use a single structured prediction task to combine all tasks. In this way, they hope to reveal the patterns of the correlation among attributes. However, the output space of the SNE is much larger than the conventional multiple task learning method. So SNE is not suitable if the scale of input data sources is limited, or it will lead to overfitting.

Recently, Kim et al. propose a new multi-task method to predict age, gender, and marital status from people's transaction records. Though this is not a SEA task, it is also a typical UAI task. Researchers collect the purchasing histories and user attributes of 56 thousand users. The input data is quite similar to SNE [96]. Compared with SNE, [54] transforms shared embeddings into task-specific embedding and detects more important signals with an attention mechanism. The results show that the attention mechanism not only increases the performance but also help to interpret how customers' attributes relate to different items. ETNA simply use the initial embeddings of items as input, which is also not sufficient for limited input data sources. Different from these works, in chapter 4, we propose to utilize second-order feature interactions to improve the performance for limited basic features.

2.3 Collaborative Filtering Recommender System

In 5, we propose a multi-task GCN-based method, which deals with missing attributes by combining recommending and UAI tasks together. The related works which most relevant to our work mainly can be categorized as three main types: 1) GCN-based CF algorithms, 2) Attribute-enhanced CF algorithms; 3) Multi-task Recommendation methods.

2.3.1 GCN-based CF algorithms

Collaborative Filtering (CF) prevalent technique in modern recommender systems [79, 27]. CF methods mainly rely on the user-item interaction data for the recommendation. User attributes can help to increase its performance if the interaction data is too sparse. Recently, inspired by the success of graph convolution network (GCN) [56] on the graph structure, researchers begin to adapt GCN to the user-item interaction graph, capturing CF signals in high-hop neighbors for recommendation [98, 111, 101]. Many CF algorithms only leverage user-item interaction data and ignore user attributes. They are referred to as pure CF algorithms in this book. Next, we will introduce some representative GCN-based CF algorithms. Most of these GCN-based CF algorithms are pure CF algorithms. They only focus on the user-item interaction data, ignoring the help of user attributes.

In [65], Monti et al. present the first GCN-based method for recommender systems, named sRGCNN. In this approach, GCN is operated on a user-user and an item-item graph to generate user and item embeddings. The user and item representations are learned iteratively using recurrent neural networks. In the end, the objective loss function of GCN and MF are combined to train sRGCNN model. sRGCNN show the potential and idea of applying GCN in recommender system. However, it did not consider the direct relationship between users and items. And the RNN-based iterative computation maybe not efficient for large-scale datasets [11].

Berg et al. propose GCMC[11], which is the first attempt to directly apply GCN on the user-item rating graph. GCMC uses one graph convolution layer to build user and item embeddings. The embeddings are updated by message propagation on the bipartite user-item interaction graph. The experiment results demonstrate that GCN-based CF is competitive with other state-of-the-art CF methods. GCMC utilizes one convolutional layer, so it actually only model the direct connections between users and items. The indirect connections, such as user-item-user directions or item-user-item directions are ignored. This will limit the representation power of GCN, which may lead to sub-optimal recommendation results.

PinSage [106], presented by Ying et al., is an improved version of GraphSAGE [38] for industrial-level large user-item graphs. In industrial scenarios, the scales of datasets are much larger than benchmarks commonly-used by academic researchers. Most of the previous GCN-based recommender systems need to be trained on the full user-item graph. Therefore the resource cost becomes unbearable when the graph has billions of nodes. Researchers present a novel sampling method for random walk based on the importance of neighborhood. PinSage leverage efficient random walks and two graph convolution layers on the item-item graph to generate item embeddings. The experiment results on Pinterest image dataset (7.5 billion training samples) show a significant improvement in recommendation performance. It is the first work that demonstrates that graph convolutional methods can be effectively applied in a production recommender system.

To improve the prediction performance in cold-start scenarios, Zhang et al. propose a stacked and reconstructed GCN, STAR-GCN [108]. Cold-start problems refer to a scenario that new users (or new items) do not have any interaction with existing items (or new users). STAR-GCN uses the masking technique to alleviate the cold start problem. It masks part of existing nodes to simulate the new user/item node for the GCN model. Besides, researchers also discover a training label leakage issue in GCN-based models implementation. The experiment results demonstrate the efficiency of STAR-GCN in cold-start scenarios. However, the ask-to-rate technique of Star-GCN might do not apply to real-world cold-start scenarios.

Wang et al. introduce NGCF to exploit the high-order connectivity from user-item interactions[98]. GC-MC only leverages the low-order connectivity between user and item. High order connectivity denotes the multi-hop path in bipartite user-item interaction map. This high-order connectivity contains rich semantics of collaborative signals, which can reveal the high-order similarity of users and items. Researchers propose a new embedding propagation layer to encodes the high-order connectivity in user and item embeddings. The experiment results on three real-world datasets shows that NGCF significantly outperforms PinSage and GCMC, which mainly consider the low order user-item relationships.

Recently, He et al. propose LightGCN, a simplified version of NGCF with better prediction performance [45]. The researchers argue that some designs in NGCF are heavy and redundant. NGCF inherits many unnecessary operations from GCN. For example, the nonlinear feature transformation is important for GCN because GCN needs to process nodes with rich attributes. However, NGCF does not consider attributes. The user or item node has no semantics in NGCF. So the nonlinear feature transformation only increases the difficulty in the training model in NGCF. By simplifying feature transformation and non-linear activation in GCN layers, LightGCN outperforms NGCF, and achieves state-of-the-art performance.

Though these GCN-based CF algorithms perform well in general scenarios, they may encounter the interaction sparsity problem. Leveraging attributes has been a successful way to help CF methods to deal with the interaction sparsity problem.

2.3.2 Attribute-enhanced Recommendation

Though CF provides a universal solution for recommendation, its performance could be affected when user-item interactions are too sparse. To alleviate this problem, researchers have developed a few attribute-enhanced CF algorithms to effectively integrate user/item attributes into user preference prediction [75, 44, 11]. Like pure CF methods, these attribute-enhanced CF methods can still get results without any attributes, though they are designed to utilize attributes. These attribute-enhanced CF can be seen as typical UAE tasks. Next, we will introduce some representative attribute-enhanced CF algorithms.

Rendle et al. propose Factorization Machine (FM) [75], which can model the second-order feature interactions to predict the interaction labels. FM first converts all information included user/item ID and attributes related to interaction to a feature vector via multi-hot encoding. Then FM estimates the target by modeling all interactions between each pair of features via factorized interaction parameters. FM is famous for its generality. It is a general predictor that can process any real-valued feature vector for supervised learning. Though considered to be one of the most effective embedding methods for sparse data, FM is essentially a multivariate linear model. Inspired by the success of deep learning, researchers try to improve the non-linear expressive power of FM through deep neural networks.

To achieve both memorization and generalization in recommender systems, Cheng et al from Google propose Wide Deep learning framework [21]. Memorization means the tasks to find the co-occurrence of features which already occurred in the past. Generalization refers to the tasks to generate new feature combinations that have never occurred in the past. Both existing FM or DNN methods are good at generalizing new features (including ID, attributes, and other features like time) combinations without manual feature engineering. However, if the original user-item interaction dataset is too sparse, FM or DNN may be hard to find effective feature combinations for users with special preferences. To capture these rare preferences (exception rules), researchers propose to manually construct new features by combining multiple predictor variables, i.e. cross features. A wide linear model is designed to memorize these cross features. A DNN model is used to generalize new feature combinations for frequent preference. Therefore, Wide & Deep model is good at both memorization and generalization. The experiment results on a large-scale industrial dataset show that the Wide & Deep learning framework significantly outperforms previous models which consider either memorization or generation.

Though top scientists from companies like Google are very good at manually designing feature combinations, the cost is usually unbearable for ordinary researchers because of the heavy engineering efforts and the high requirement for domain knowledge. FM can automatically generate feature combinations. So a lot of efforts are put to improve the performance of FM. He et al. argue that the performance of FM is be limited because it only models linear and second-order feature combinations. He et al. propose NFM [44] proposes to effectively model higher-order and non-linear interactions among attributes. Through designing a new operation in neural network modeling — Bilinear Interaction pooling — the researchers combine FM into the neural network framework. The shallow linear FM is deepened by several non-linear neural network layers above the Bi-Interaction layer, to model higher-order and non-linear feature interactions. The results show that NFM effectively improves FM's expressive power.

Besides static attributes, users' sequential behaviors are also important in some time or location-sensitive scenarios. These behavior data indicates users' dynamic and evolving interests. For example, people's purchasing behaviors are different as time evolves or location changes. Zhou et al. from Alibaba propose Deep Interest Network (DIN) [113] to combine users' historical behaviors w.r.t. the target item and user/item attributes to learn user/item representation. Previous works usually use fixed-length representation. Zhou et al. argue that this could be a bottleneck for capturing the diversity of user interests. DIN utilizes an adaptive representation vector for user interest which varies over different items to improve the expressive power. The test results of deployment in the production environment of Alibaba show that DIN outperforms previous sequential-based or non-sequential recommender systems.

Several GCN-based methods like GCMC and Star-GCN also consider user/item attributes as input to improve recommendation performance. For example, GCMC uses a multi-layer perceptron to model user/item attributes, which is separate from GCN. These methods have shown remarkable performance in sparse or cold-start scenarios, indicating that integrating the attributes features and user-item interactions are helpful.

However, these methods, especially those based on GCN, seldom discuss one problem: missing attributes, which is quite common in real-world scenarios. An easy way is to fill the missing features with the most frequent feature values, zeros, or unknown tags. These substitute values make these attribute-enhanced methods able to run and generate outputs. However, if the missing rate is too high and the attributes are quite important to recommendation performance, then simple substitutes maybe not enough.

CC-CC [80] tries to tackle feature missing problems by random feature sampling and adaptive feature sampling strategies. However, they mainly focus on the lower missing rates (10%-30%). Its performance will be also affected if the missing rate is too high. Because there are not enough features to learn proper sampling strategies. In 5, we use a multi-task learning GCN-based method to estimate attributes for recommending task, which can alleviate the problems caused by high missing ratio of attributes.

2.3.3 Multi-Task Learning for Recommender Systems

As discussed in chapter 2.2.2, MTL is a learning paradigm in machine learning, which aims to leverage shared information on multiple related tasks to improve the performance of all the tasks[109]. MTL methods in the field of UAI mainly refer to predicting multiple attributes. These tasks are all UAI tasks, which are similar to each other except for the label data. In the field of recommendation, there are also MTL methods that aim to find the relationship among several recommending tasks, like [103]. However, there is also another problem: researchers often need to utilize the relationship between recommendation and tasks which are not recommendation, like natural language processing, computer vision, etc. And researchers often care more about the results of recommendation instead of all tasks. In this way, the non-recommendation tasks can be seen as auxiliary tasks which aim to help improve the performance of the main task, i.e., recommendation. Next, we will show some successful applications of deep learning-based MTL in recommendations in these two scenarios.

Shen et al. present a multi-task method DINOP to tackle Sales Predictions for Online Promotions (SPOP) problems[103]. In this work, SPOP means a group of sales-related forecast tasks for the promotion day, which is an important problem in the real commercial environment. On the promotion days like Black Friday, the e-commercial platforms need to predict several targets in advance, including gross merchandise volume (GMV), sales volume (SV), best-selling products (BSP), etc. GMV indicates the total income for goods sold during a certain period, which is one of the most important values for the promotion day. SV indicates the total amount of a kind of commodity. BSP reveals the most popular items. Though different, all these targets belong to recommending tasks. And part of their input features is the same as each other. The researchers argue that time and resources could be wasted in designing and training models for numerous new tasks sharing a part of the same features. Besides, the input features are also insufficient because the same promotion days only occur one time in one year, while the corresponding data is usually only kept in the database for a short time (e.g., 2 years Alibaba). MTL can help to alleviate the insufficient data problem. Researchers propose Deep Item Network for Online Promotions (DINOP), a multi-task learning method to learn general representation among several SPOP problems. The experiments on a large- scale industry data set validate the effectiveness and efficiency of DINOP.

Lu et al. present a multi-task learning framework that can recommend items to users and give an explanation for why recommending the items at the same time [59]. In previous works, researchers already find out that the text content of users' reviews can be used to enhance the performance of recommendation. In this work, the researchers use adversarial sequence-to-sequence learning techniques to generate textual-based reviews of users as an explanation. The generated reviews not only provide a human-friendly explanation for recommendation but also play as auxiliary information to improve the performance recommendation. The experiment results on real-world datasets demonstrate the joint training model significantly outperforms many existing single-task-learning recommendation methods. And the generated reviews are also much closer to the ground-truth data than previous single-task-learning approaches.

In [32], Gao et al. present Neural Multi-Task Recommendation (NMTR) to model users' multi-behavior data. Previous recommender systems usually only utilize one kind of user behavior data. For example, in E-commerce, researchers often only discuss purchasing behavior data. Actually, the other kinds of user behavior data are also important, like users' views, clicks, and adding to the shopping cart or collection. In NMTR, the researchers pay special attention to the converting order among different behaviors (e.g., view or click before purchasing). NMTR jointly model the interaction of all behaviors and the cascading relationship among each kind of behavior. The experiments demonstrate that NMTR outperforms existing recommender systems on multi-behavior data.

Different from these works, in 5 we focus on leveraging recent advances in GCN to exploit the commonality between recommending and UAI in the multi-task learning (MTL) approach. In this approach, rating prediction is the main task and UAI is the auxiliary task. Our main purpose is to extend UAI which only care about its own performance to a framework which can directly help its downstream UAE tasks, e.g., recommendation.

Chapter 3

Single-Attribute-Level Problem: Enabling Human Mobility for Socioeconomic Status Estimation

In this chapter, we main discuss a Single-Attribute-Level Problem of UAI: enabling new type of user-generated data sources for attribute inference. As a case study, we focus on introducing Smart Card Data (SCD), which records the temporal and spatial mobility behavior of a large population of users, into individual-level SES prediction.

Contents

3.1 Introduction

Socioeconomic Status (SES) is an economically and sociologically combined overall measure of an individual or family, typically based on income level, education level, and occupation [16, 84]. SES reflects the corresponding a person's social and economic rank in society. And it is typically divided into three levels (high, middle, and low)[16]. An individual with a higher SES means he/she earns more, has a better job or higher education than those with a lower SES. SES nowadays plays an important role in many areas like sociology, economics, public administration, and education. It can help governments to design and evaluate social policies, especially for welfare policy. Recently, companies become more and more interested in assessing people's SES because it is a valuable demographic feature to many emerging applications, such as customized marketing, personalized recommendation, and precise advertisement [91, 19, 51, 102]. Especially, in personal credit rating, SES is an important factor that helps online banks (e.g., Lending Club[1]) to decide the volume of loans they will lend to an individual [91].

Given its importance, various approaches have been developed to measure SES, most of which need to collect at least one kind of the following information: individual income, education or occupation [16], typically through real-world contacts with the individuals under investigation. For a large-scale investigation covering millions of people, it is usually conducted through household interviews by National Statistical Institutes. Some researchers or professional investigation companies also try to collect SES information through methods like online questionnaires or telephone surveys. However, most of them can only cover a small group of people. Although traditional methods can get very detailed information, the investigators usually publish regional-level statistics instead of individual SES information (which is much more important to many companies). Also, the time gap between two successive

---

[1]lendingclub.com, one of the largest peer-to-peer lending platform.

large-scale surveys could be very long, which may even be several years. If companies decide to collect SES by themselves, they find that the cost is unbearable and many citizens are also quite reluctant to expose their real income or job information. Even governments of some developing countries are also facing the same problem [12].

Due to the prohibitive costs and time required to collect large-scale individual-level SES information, researchers try to estimate individual-level SES using some easily accessible user-generated data sources like online social networks [72, 73, 58], Although most existing big data-based methods can only get a rough income level (low, middle, high) of people, they are still valuable to many companies and researcher, owing to their substantially lower cost and time in estimating SES for a large user population. Further, to better support targeted applications it becomes necessary to improve the accuracy of big data-based SES estimation via better algorithms or different data sources with lower costs or privacy concerns. This chapter attempts to answer the following question: *Can SES be roughly estimated based on human mobility-related data alone?*

Data-based SES estimation methods are actually based on an observation that different SES levels of people may have different lifestyles. Lifestyle depicts typical routine lives of people. Large-scale human mobility data like smart card data (SCD) or online check-in data can act as an approximation for human lifestyle. Previous methods [87, 105, 12] based on cellphone discussed some general statistical mobility features. However, these features are simply complemented to specific cellphone features like the numbers of calls and telephone fares. These mobility features may not be enough for organizations (e.g., public transit agencies) which only have human mobility data. In this chapter, we study whether we can get a satisfactory estimation of user-level SES when we only get users' mobility data.

As a case of mobility data source, we take SCD generated by smart card automated fare collection systems, which are now widely used by public transit agencies. Essentially, SCD is administrated by a city municipality and records a large number of individual-level, time-stamped and geo-tagged trip data of its citizens [9, 64]. Although a large and growing body of work has studied SCD in different contexts, little attention has been paid to estimate SES based on SCD. We develop S2S (*Smartcard* to *SES*), a method for estimating SES based on SCD and other related public information. The main challenges in designing S2S are:

- Designing effective features related to SES based on smart card data.

- Designing a model which can utilize different types of features to improve the performance of estimation.

To the best of our knowledge, this work is the first attempt to estimate user-level SES using SCD data. Our main contribution is summarized as follows.

- We propose a deep neural network (DNN)-based learning approach (S2S), which considers both temporal-sequential features and general statistical features of human mobility. Especially, the sequential aspects are considered in S2S, representing more salient nature of an individual's behavior in socioeconomic context than traditional general statistical features.

- We evaluated our approach using actual large-scale SCD data of totally 7,919,137 cards of Shanghai City for 16 consecutive days. The results demonstrate our approach significantly outperforms several baselines.

The rest of this chapter is structured as follows. Section 3.2 introduces the datasets. Section 3.3 discusses the features. The S2S model

Table 3.1: Subway Record Example

ID	Date	Time	Station Name	Fare
1000019	2015/04/02	17:01:05	station A	0.0
1000019	2015/04/02	17:35:49	station B	4.0
1000039	2015/04/06	18:03:04	station C	0.0
1000039	2015/04/06	18:17:49	station D	2.0

is proposed in Section 3.4. Experimental results on Shanghai SCD are presented in Section 3.5. The chapter is concluded in Section 3.6 with a brief discussion of limitations and directions of future research.

3.2 Datasets

3.2.1 Data Collection

We exploit three related datasets in this work: smart card, POI and housing price. We describe them respectively below.

Smart card: The smart card dataset is opened by the Shanghai Open Data Applications contest. The dataset contains all the subway records in Shanghai between April 1st and April 16th, 2015. The example format of a subway record is shown in Table 3.1. One single subway trip consists of two successive records. The first one is created when the user gets into the boarding subway station and begin to travel in the subway system. The second record is created when the user gets out of alighting station. If the fare is 0.0, then the user is getting aboard a metro train, or they are getting off. There are 7,919,137 IDs which can be correctly recognized after data cleaning. When users apply for a smart card in Shanghai, they do not need to provide any personal information. So IDs do not have any relationship with real-world identification, avoiding possible privacy leakage.

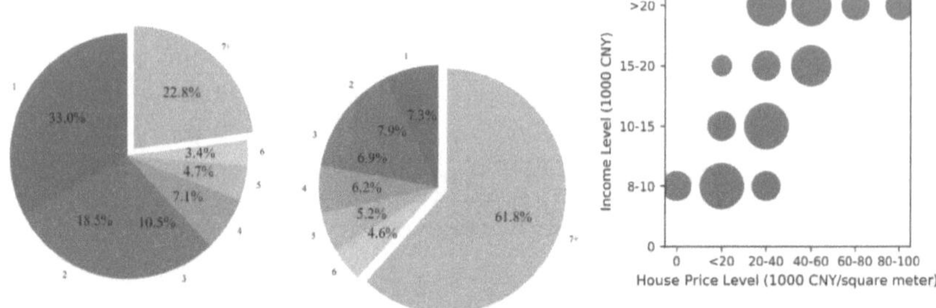

<table>
<tr>
<td>

Figure 3.1: User distribution: only 22.8% are frequent user who take subway more than 7 days

</td>
<td>

Figure 3.2: Trip Distribution: frequent user takes more than 60% subway trips

</td>
<td>

Figure 3.3: The relationship between house price and monthly income: larger size means more people.

</td>
</tr>
</table>

POI: POI dataset of Shanghai is crawled based on GaoDe Map API Service[2]. The categories include Public Facility, Domestic services, Education, Business Residence, Hospital, Hotel, Car services, Sport&Leisure, Scenery, Restaurant, Public Transportation and Financial Services.

Housing price: Housing price dataset is crawled from Lianjia.com [3] website, which records the house prices and location information of most apartments/houses for selling in Shanghai. We crawl the average housing prices of all communities (a community usually includes many similar houses in one area).

3.2.2 Ground Truth Construction

There are two problems in Ground Truth Construction. First, some users may only use subway for very few times (1 time) during all 16 days. We need to filter users with too less records. Second, there is no

---

[2]lbs.amap.com, one of the major online map providers in China

[3]sh.lianjia.com, one of the biggest real estate agency service providers in China.

SES information for millions of smart card holders. Automated fare collection (AFC) systems are just designed for billing purpose, so they do not collect socio-demographic information of the card holders in most cities. This Shanghai dataset (nearly 8 million smart cards) is also totally anonymous without any SES-related information, such as occupation, education and income. And we cannot manually relate smart card IDs with volunteer users because IDs have been hashed before opened for researchers. So it is hard to get actual SES label for each ID. We need to find a reasonable SES label for millions of users.

Selecting Frequent Users

As shown in Fig 3.1, although there are millions of subway users, most of them take very few subway trips. the largest group of users (33.04%) only takes subway in 1 day. More than half of people takes subway in less than 2 days. Only 22.8% of users have subway trips more than 7 days. And we also checked the trip numbers, 36.9% of users only took 1 trip. These infrequent users just use the subway occasionally. Subway is not an important transportation method for them. Their mobility data in subway system may be just a random and unimportant action in their regular life. In this work, we focus on users who have taken subways for at least 7 days. In this way, we selected about 700 thousands frequent users.

Though the number of frequent users is much smaller than infrequent users, the total number of trips they take is much more than the others. As shown in Fig 3.2, more than 60.1% of trips are taken by frequent users, who take subway more than 7 days.

Labeling Frequent Users

Getting SES label is a common problem when estimating SES for a large number of people[12, 85, 28]. Many works use the housing price of people's living place as a proxy to represent people's possible SES [64, 105, 49, 39, 52, 34, 24]. And [105] finds out that the average housing price and the income level at the corresponding area are strongly correlated (0.88). As shown in Fig 3.3, we also held an online survey [4], which collect 78 Shanghai inhabitants' monthly income and housing price. To protect the privacy and get more successful responses, we use income levels (e.g, 5,000-10,000 CNY) instead of accurate numbers. So some answers may overlap in Fig. 3. We use the size of the bubble to show the overlapped number. Bigger bubble means more same answers. We can see, the income level generally increases along with the housing price. Pearson's correlation is 0.68. The correlation is not so strong as in [105]. This may be partially caused by the phenomenon in China that some low-income young people buy high-priced houses with the help of their families. However, high family income may still also be a "bonus" to people's SES. So in general, we think housing price is a good indicator of the people' SES. Thus in this work, we use people's house price as an approximation of frequent users' SES. First, we use the method in [114] to find frequent users' home station (the station nearest to their home). Then, we select the communities around the home stations (less than 2 km), to calculate the average housing price of the home station. SES is usually divided into 3 levels: high, middle and low. We divide frequent users into 3 levels based on the average housing price of their home station. There are 19.4% of users at high level (housing price $>$ 70000 CNY/m^2), 36.2% in middle level and 44.4% at low level.

3.3 Feature Engineering

3.3.1 Overview

A user's smart card records can be seen as a list of tuples $\{(s_1, t_1, ao_1), (s_2, t$ s_i and t_i denote the subway station and the time of the i-th record. ao_i denotes whether the user is getting aboard and off at i-th record. Given users' smart card records, we aim to estimate users' SES levels. The overall research design is shown in Fig 3.4. One of the key challenges is

---

[4]http://wj.qq.com/s2/3598293/4053/

feature engineering. We mainly utilize two types of features in this work: general statistical features and temporal-sequential feature. General features (shorten form of general statistical features) usually consider the statistical features of a user's whole mobility data. They have been discussed by previous works like [87, 69, 105]. However, previous papers largely neglect the temporal and function information related to each station, which will be discussed in following section.

3.3.2 General Feature

F_{rg}, **Radius of Gyration**

F_{rg} is defined as follows:

$$F_{rg} = \frac{\sum_{i=1}^{n} distance(\vec{s_i}, \vec{s_c})}{n} \tag{3.1}$$

Here, $\vec{s_i}$ denotes the location (latitude and longitude coordinates) of s_i. $\vec{s_c} = \sum \vec{s_i}/n$ denotes the geographic center of all s_i. $distance$ is the geographic distance between two locations. A large value of F_{rg} indicates the user mobilize in a large area.

F_{krg}, **K-Radius of Gyration**

Let $count(i)$ be a counting function, which is equal to the number of s_i in a user's whole mobility record. A large value of $count(i)$ means the user often visit the subway station s_i. F_{krg} is a radius of gyration calculated using only top k visited stations. [69] proposed it to measure how a user's top k stations determine his/her radius of gyration. F_{krg} is defined as:

$$F_{krg} = \frac{\sum_{i=1}^{k}(count(i) \cdot distance(\vec{s_i}, \vec{s_c}))}{\sum_{i=1}^{k} count(i)} \tag{3.2}$$

The aim of F_{krg} is to find out returners and explorers. [69] suggested that, k-returners are those whose $F_{krg} \geq F_{rg}/2$ and k-explorers are

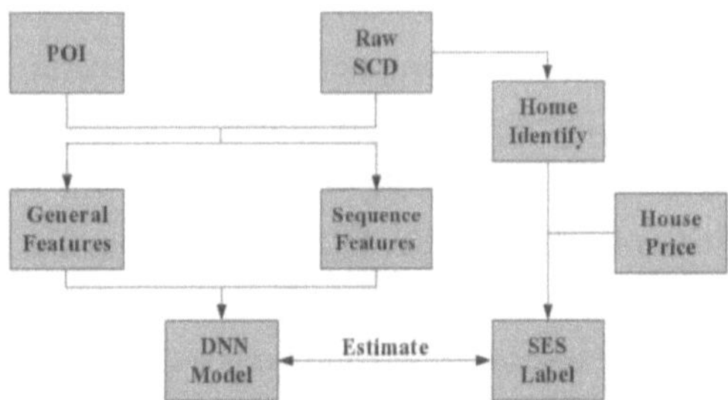

Figure 3.4: Overall research design

those for whom $F_{krg} < F_{rg}/2$. We can simply think that k-returners are those who tend to spend most of the time between k the most important locations, while k-explorers are those whose activity space cannot be well described by only k top locations. And in this work, we set $k = 2$. In this way, 2-returners are likely to be a common commuter between home and working place.

F_{nds}, **Number of Different Stations**

F_{nds} is defined as follows:

$$F_{nds} = |set(s_1, s_2, \ldots, s_n)| \tag{3.3}$$

F_{nds} measures the total number of different stations visited by a user during all 16 days. A larger value of F_{nds} means that the users tend to visit more different subway stations.

F_{ae}, **Activity Entropy**

Given a vector $\{p_1, p_2, \ldots, p_{\bar{n}}\}$, where $\bar{n} = F_{nds}$ and $p_i = \frac{count(i)}{\sum_{i=1}^{n} count(i)}$. p_i denotes the proportion of visiting numbers of station s_i, the activity entropy is calculated as:

$$F_{ae} = -\sum_{i=1}^{\bar{n}} p_i \log(p_i) \tag{3.4}$$

A large value of F_{ae} means that the spatial diversity of a user's daily activities is high.

F_{td}, Travel Diversity

Travel diversity measures the regularity of a user's movements among his/her subway stations. We define an origin-destination trips as a trip between two consecutive stations. Let E denote all the possible origin-destination pairs (without considering direction) extracted from set $(s_1, s_2, \ldots, s_{\overline{n}})$, all stations a user visit. Then the travel diversity is defined as:

$$F_{td} = - \sum_{i \in E} p'_i \log(p'_i) \tag{3.5}$$

where p'_i is the probability of observing a trip between the i-th origin-destination pair. A large value of F_{td} means that a user's tend to travel between quite different origin stations and destination stations.

3.3.3 Sequence Feature

People may tend to follow regular and stable patterns during their everyday lives. And people in different SES-level may visit different places and have different commute schedules. For example, cleaners usually need to go to company earlier while IT engineers may have to work at company until very late at night. Here we use sequence feature (shorten form of temporal-sequential feature) to describe these phenomenons.

We divide all 16 days into 1536 (16x24x4) time bins by every 15 minutes. For each time bins, we need to find the location where a user stay, and calculate a feature vector based on the location. Given that a user's sequence feature is $\{X_1, X_2, \ldots, X_i, \ldots, X_N\}$, where $N = 1536$ and X_i denote the feature vectors of location at the i-th time bins. X_i consists of three kinds of features: the ID of time bins ($timeID$, from 0 to 1535), function of station for most citizens ($F_{fm}, \{residential, entertainment, working, transfe$ and function of station for current user ($F_{fu}, \{home, work, others, transfer\}$).

To find the location where a user stay, first we take the stations as the location of the corresponding time bins. For example, if during the first time bins, a user get aboard on station A, then we take station A as the user's location of the first time bins.

Then for time bins which there is no corresponding station, we use following method to find their approximate locations:

① Among the time bins with a station location, find out those when the user is getting aboard and the others when the user is getting off, based on ao_i. The former time bins are denoted as $T_{aboard} = \{t_{a1}, t_{a2}, \ldots, t_{ai} \ldots\}$. The latter time bins are denoted as $T_{off} = \{t_{o1}, t_{o2}, \ldots, t_{o}$

② If a series of time bins are between two consecutive stations, t_{oj} and t_{ak}(the first for getting off and the second for getting aboard), the locations of the first half time bins are the station of t_{aj} while the second half are the station of t_{ak}.

③ If a series of time bins are between two consecutive stations, t_{al} and t_{om}(the first for getting aboard and the second for getting off), we do not need to find their locations. The detail of how to calculate the feature vectors for these time bins will be discussed in following sections.

④ For the time bins before t_{a1}, the locations are the station of t_{a1}.

⑤ For the time bins after last getting off station (i.e, t_{oN}), the locations are the station of t_{oN}.

F_{fm}, Function of station for Most citizens

The step of urbanization leads to different functional regions in a city, e.g., residential areas, business districts, and entertainment areas [107]. People show in the different functional areas may have different social attributes. For example, housewives may mainly stay inside residential areas while regular office worker may travel between the residential area and business districts during the weekday. And different kinds of people may spend different time in some special functional regions. For example, a rich family may spend more time in entertainment areas during the weekend than an ordinary family. Here we use two features called F_{fm} to describe this phenomenon.

Here we explain how to determine the function for each subway station. There are different functional regions in one city, supporting different needs of people's urban lives. And similarly, each subway station also has a different function. People tend to use the subway station which is nearest to their starting location and ending location. For example, if a subway station is inside a residential area, then most people using this subway should be the people who live near this station. During the weekday, most users of this subway station would get into the subway in the morning to go to work and get out of the station in the evening to go back home. On the other hand, if a subway station is inside a work area, surrounded by a lot of companies, then most people using this subway should be the people who work near this station. During the weekday, most users of this subway station would get out the subway in the morning to go to work and get into the station in the evening to go back home. So the function of one subway station is actually the function of the area near it.

In this work, we use the same method in [107] to divide all Shanghai subway station into 3 kinds: residential, entertainment and work. This method needs to consider the human mobility and poi data of each station. The distribution of function stations is shown in Figure 3.5. The blue points represent residential stations, the red points represent entertainment stations and the yellow points represent work stations.

For most X_i, F_{fm} is "residential", "entertainment" or "working". However, if X_i is between two consecutive stations, t_{al} and t_{om}(the first for getting aboard and the second for getting off), F_{fm} is "transfer". It means the user is traveling from one function area to another function area.

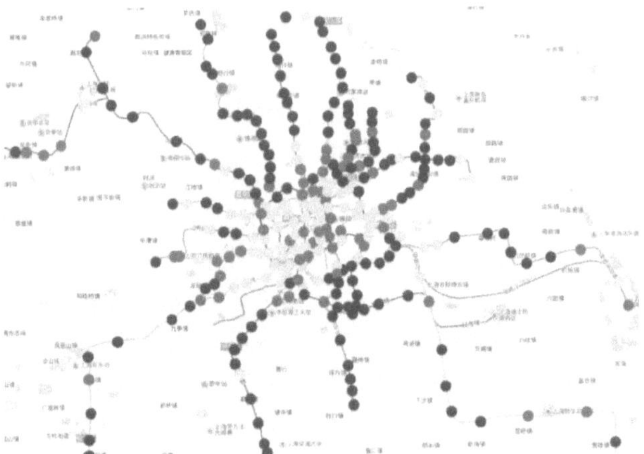

Figure 3.5: Function Station Distribution in Shanghai: blue = residential, red = entertainment, yellow = work, the lines are the subway lines, the points are the subway stations.

F_{fu}, **Function of station for current user**

For some users, the function of a specific station may be different from most users. For example, someone may work in a supermarket in a living area. Though for most people, the station is a "residential" station. However, for this person, the station is more like a "working" station.

In this work, we use the same method in [107] to divide a user's stations into 3 kinds: "home", "work" and "others". For most X_i, F_{fu} is "home", "work" or "others". However, if X_i is between two consecutive stations, t_{al} and t_{om}(the first for getting aboard and the second for getting off), F_{fu} is "transfer".

3.4 S2S Model

The goal of the proposed model is to estimate a user's SES level, denoted as Y_{uid}, where uid is the id of a smart card user. Fig. 3.6 shows the architecture of the proposed model, which is comprised of two major components. The sequential component processes sequence features and outputs Y_s. The general component processes general feature and outputs Y_g. Y_s and Y_g are fused and fed into the softmax layer to estimate the SES level of input user.

3.4.1 Sequential Component

People of different SES level may have different lifestyles, like visiting different places and having different commute schedules. We need to capture the temporal dependence of people's mobility. The recurrent neural network (RNN) is an artificial neural network which is widely used for capturing the temporal dependency in sequential learning, such as the natural language processing and speech recognition [41]. When processing the current time step in the sequence, it updates its memory (also called hidden state) according to the current input and the previous hidden state. The output of the recurrent neural network is the hidden state sequence at all the time steps in the sequence. The sequential feature we design considered the transition of different function stations, which can be effectively handled by RNN. Sequential component is composed of an embedding layer, a single RNN layer, and two fully-connected layers, as shown in Fig. 6. In this work, we denote the feature at time bin i as $X_i = (timeID, F_{fm}, F_{fu})$. In our experiments, RNN performs not so well in processing the long time bins due to vanishing gradient and exploding gradient problems. Therefore, instead of the RNN layer, we adopt the Long Short-Term Memory (LSTM) [33] layers. In short, LSTM adds an input gate and a forget gate to alleviate the gradient vanishing/exploding problem.

X_i is fed into an embedding layer first. Because $timeID$, F_{fm} and F_{fu} are both categorical values which can not feed to Recurrent Neural Network (RNN) layer directly [31]. The embedding layer transform $timeID$, F_{fm} and F_{fu} into three low-dimensional real vectors ($timeID^e$, F_{fm}^e and F_{fu}^e), respectively. The $timeID^e$, F_{fm}^e and F_{fu}^e are concatenated to get X_i^e. X_i^e is fed into the LSTM layer, which output a hidden state h_i. We concatenate all of the hidden state frag-

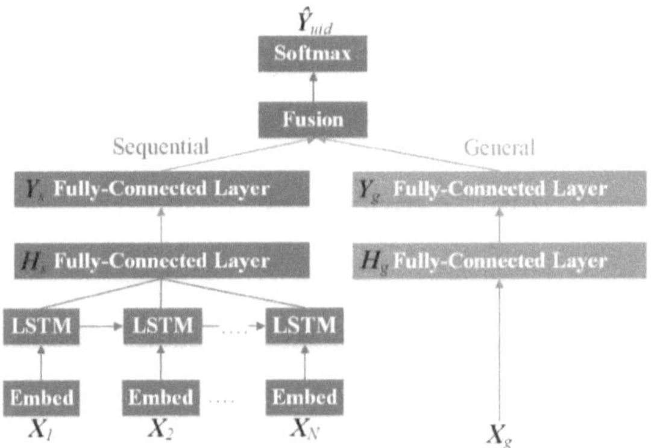

Figure 3.6: Model Architecture

ment $[\boldsymbol{h}_1, \boldsymbol{h}_2, \ldots, \boldsymbol{h}_i, \ldots, \boldsymbol{h}_N]$ as $\boldsymbol{H}_N$. Then $\boldsymbol{H}_N$ is fed into the fully-connected layer as:

$$\boldsymbol{H}_s = ReLU(\boldsymbol{W}_{hs}\boldsymbol{H}_N + \boldsymbol{b}_{hs}) \tag{3.6}$$

$\boldsymbol{H}_s$ is then fed through the fully-connected layer to output the $\boldsymbol{Y}_s$, defined as:

$$\boldsymbol{Y}_s = \tanh(\boldsymbol{W}_s\boldsymbol{H}_s + \boldsymbol{b}_s) \tag{3.7}$$

where $\boldsymbol{W}_{hs}, \boldsymbol{b}_{hs}, \boldsymbol{W}_s$ and $\boldsymbol{b}_s$ are the learnable parameter matrices used in the fully-connected layers.

3.4.2 The Structure of General Component

Besides the sequence feature, the general mobility feature may also reflect a part of lifestyles. We discussed these features and their possible relationship with SES-level in Section IV. We stack two fully-connected layers to model the general factors that affect SES. $\boldsymbol{X}_g = [F_{rg}, F_{krg}, F_{nds}, F_{ae}]$. The first layer processes the feature vector $\boldsymbol{X}_g$ and outputs a hidden state $\boldsymbol{H}_g$:

$$\boldsymbol{H}_g = ReLU(\boldsymbol{W}_{hg}\boldsymbol{X}_g + \boldsymbol{b}_{hg}) \tag{3.8}$$

Then $\boldsymbol{H}_g$ is fed into the second layer and get the output of the general Component $\boldsymbol{Y}_g$:

$$\boldsymbol{Y}_g = \tanh(\boldsymbol{W}_g \boldsymbol{H}_g + \boldsymbol{b}_g) \tag{3.9}$$

where $\boldsymbol{W}_{hg}, \boldsymbol{b}_{hg}, \boldsymbol{W}_g$ and $\boldsymbol{b}_g$ are the learnable parameter matrices used in the fully-connected layers.

3.4.3 Fusion and Training

We here combine the output of the two components as shown in Fig. 6. The fusion layer assigns the weights to two components. Finally, the softmax layer estimates the SES level of a user denoted by $\hat{\boldsymbol{Y}}_{uid}$. $\hat{\boldsymbol{Y}}_{uid}$ is defined as:

$$\hat{\boldsymbol{Y}}_{uid} = Softmax(\boldsymbol{V}_s \circ \boldsymbol{Y}_s + \boldsymbol{V}_g \circ \boldsymbol{Y}_g) \tag{3.10}$$

where $\circ$ is element-wise multiplication, V_s and V_g are the learnable parameters that adjust the contribution of sequence and general features to $\hat{\boldsymbol{Y}}_{uid}$. The model can be trained by minimizing the cross-entropy between the ground truth $\boldsymbol{Y}_{uid}$ and the estimated SES level $\hat{\boldsymbol{Y}}_{uid}$:

$$\zeta(\theta) = -\boldsymbol{Y}_{uid}^U \log \hat{\boldsymbol{Y}}_{uid}^U \tag{3.11}$$

where θ are all learnable parameters of S2S model and U means the user number for training. We first construct the training dataset from a part of users' actual SES level and corresponding features. Then, S2S model is trained via back-propagation and Adam [55].

3.5 EXPERIMENTS

3.5.1 Settings

The details of datasets and ground truth are already introduced in Section III. Finally, We picked 729,859 users who take the subway for at least 7 days (during 16 days). These users are divided into 3 SES levels:

high, middle and low. 80% of picked users are for training and 20% for testing. The results are mainly measured by classification precision, recall, and F1-score.

To the best of our knowledge, there exists no model directly estimating SES from users' SCD. We use the following baselines to test the effectiveness of our model:

① **Random Guess** just randomly classifies the user to an SES label.

② **Spatiality, Temporality, and Location Knowledge (STL).** This method predicts twitter users' demographics based on their online check-ins [112]. Online check-ins are another kind of mobility data. They are uploaded to online social networks by people to show where and when they are. STL organizes users' check-ins into a three-way tensor representing features based on spatial, temporal and location information (e.g, location category, keywords, and reviews of a POI). Then a support vector machine (SVM) is trained for estimate users' demographics (e.g., gender, blood type). We treat station records as users' check-ins when using STL. However, we have to omit some location information like reviews. Because subway stations just do not these kinds of data.

③ **Gradient boosting decision tree (GBDT).** The gradient boosting model is famous for its outstanding performance and efficiency for estimation. The LightGBM is an open source gradient boosting library [53]. It has been widely adopted in many data mining competitions like Kaggle. We use sequence feature and general feature to train LightGBM model.

Besides the above baselines, Sequence model (**S2S-S** model) and General model (**S2S-G** model) are also tested to find out the most effective feature categories. S2S-S model only uses sequential features with sequential component. S2S-G model only uses general features with general component. We refer our method which involves both sequence and general feature as **S2S-SG**.

Parameter Setting. The main parameters of our experiment are as follows. In the embedding layer, we embed $timeID$ to R^{11}, F_{fm} to R^2 and F_{fu} to R^2. In the general component, the neuron number of two fully-connected layers are both 24 neurons. In the sequential component, the size of the hidden vector h_i is 64. In the fusion component, the size of the hidden vector Y_s is 24.

The learning rate of Adam is 0.001 and the batch size during training is 12000. Our model is implemented with Keras. We train our model on a 64-bit server with 12 CPU cores, 64GB RAM and NVIDIA 1080Ti GPU with 12G VRAM.

3.5.2 Performance Comparison

Table 3.2 shows the performance of baselines and S2S, and note the averages of 3 classes are used as the main comparison metric. From the result, we can see that all the metrics of S2S-SG performs better than all baselines, achieving 69% in precision, 67% in recall and 68% in F1-score. Table 3.3 shows the performance of S2S-SG in each SES class.

As shown in Table 3.2, STL is clearly better than Random Guess while less accurate than LightGBM. The reason why STL does not perform well on smart card dataset might be caused by two reasons. First, STL did not design features or methods specifically for SES estimation. Also, the subway station does not have one of the important information which STL relies on, i.e., people' reviews and keywords. Reviews and keywords of locations may also contain useful information about SES. However, unlike restaurants in STL, subway station did not have similar review information. LightGBM is better than STL, showing the proposed features are more suitable to estimate SES based on SCD. Lightgbm

Table 3.2: Comparison of each methods

Algorithm	Precision	Recall	F1
Random Guess	0.35	0.33	0.33
STL	0.49	0.42	0.45
LightGBM	0.58	0.57	0.58
S2S-S	0.63	0.62	0.63
S2S-G	0.53	0.51	0.52
S2S-SG	0.69	0.67	0.68

underperforms S2S-GS, likely due to the fact Lightgbm underperforms LSTM on understanding long sequential features.

We can also see that S2S-SG outperforms the other S2S models. S2S-S is clearly better than S2S-G, demonstrating the value of sequential features. And the performance of S2S-S is even better than LightGBM with full features. There may be two reasons why general statistical features are not so useful as sequential features. First, the dataset covers only 16 days. The cellphone datasets which previous works studied usually last for months. So the general feature here may be not suitable for short time. Second, general features are not good at capturing some subtle differences in people's lifestyles. For example, some high SES-level people like to go for entertainment instead of going back home after work, while some low SES-level people also visit such an area for part-time work. It is hard to distinguish them based on general features because they may all have a larger mobility area than others, like home-work commuters. However, sequential features can help in these scenarios, e.g., checking whether one goes to a station for work or for entertainment, or checking whether one is going to an entertainment area during usual working time (e.g, 9am-5pm every workday) or after work (e.g, after 8 pm). Also, people who go to entertainment areas during work time are more likely to be a service staff than a consumer.

We also manually check some error estimations. We find out that many users in high SES-Level are mislabeled as middle SES-level. This may be because most frequent SCD users are not so "rich". Actually,

Table 3.3: Performance of S2S-SG

SES-Level	Precision	Recall	F1
High	0.69	0.55	0.61
Middle	0.65	0.67	0.66
Low	0.74	0.80	0.77
Avg	0.69	0.67	0.68

most subway-frequent users are middle and low-income levels among the city's population, so their difference may not so clear. Besides, we just differ high SES-level or middle SES-level people based on their housing price (70,000 Chinese Yuan (CNY)/m^2). However, there is a large group of users who are around the 70,000 CNY/m^2. We checked their home stations. Many middle and high price-level home stations are quite near to each other. So the difference of mobility feature between them is also not so clear. It means we still need to improve the features in our future work.

3.6 Chapter Summary

This chapter examines whether people's SES can be estimated only based on their smart card mobility data. We take the Shanghai smart card data as a case study. Because individual-level income information is hard to get for millions of people, we hypothesize that people's income level is related to the house-price level of their home. In this way, we get the SES label of about 700 thousand users who frequently take subways. Mobility features and a DNN model named S2S, are proposed to estimate their SES-level. In the end, experiments show that these SCD-based features can be used to estimate the SES level (much better than random guess), wherein the sequential features are clearly better than traditional general features. This method can be used to quickly give a rough individual-level SES estimation for millions of people, when companies or researchers can only get people's mobility data.

Chapter 4

Multi-Attribute-Level Problem: Multiple Socioeconomic Attributes Estimation based on Home Location

In this chapter, we mainly discuss a Multi-Attribute-Level problem of UAI: improve the performance of multi-attribute prediction with limited input data sources. As a case study, we focus on inferring multiple socioeconomic attributes solely from users' home location.

Contents

4.1 Introduction

Inferring people's socioeconomic attributes (SEAs), such as income level, education level and occupation types, are an important problem for social computing [4]. These attributes play an important role in studies like social stratification. They also can help governments to design and evaluate social policies, especially for welfare policy. Recently, they become crucial for online service providers to offer personalized services in recommendation and advertisement [91, 19, 51, 102]. However, these attributes are hard to collect for both researchers and companies, since people are reluctant to expose their income or job information or the legal privacy framework does not allow.

Given its importance, various machine learning methods have been proposed to automatically estimate people's socioeconomic attributes (SEAs) from their cyberspace behavior [87, 12, 5, 105, 72, 73, 58]. For example, [72, 73, 58] explore how to estimate people's income or occupation based on the language patterns, topics, or even emotions in tweet content. More recently, researchers begin to get interested in inferring SEAs from peoples' physical behaviors. For instance, [96, 68] estimates people's income and education level based on how people purchase items in offline retailers.

However, home location, as fundamental user behavior, has been overlooked by most previous works for SEA inference. Previous works mainly focus on utilizing people's home location for targeted ads of local business [2], urban planning [36], location-aware recommendations [48, 88], etc. In fact, there is a common observation that SEAs such as income and occupation may be highly related to the personal home location, like in [105]. For example, richer people tend to live in areas with higher housing prices. In this chapter, we try to infer various SEAs only through people's home location and investigate how home location affects people's different SEAs.

Though important, investigating the relationship between people's SEAs and the home location is quite challenging for the following reasons. First, though datasets containing both personal SEAs and home location are critical for meaningful experiments, there are almost no open datasets including this information as far as we checked. Second, the home location itself only contains limited information for predication. Attribute prediction is hard with a limitation of input features. And personal income, occupation or education levels are complex attributes that are hard to predict even with rich human behavior data like in [104, 12].

To tackle these problems, we propose a home to SEA (*H2SEA*) method to infer people's attributes including: *personal's monthly income level*, *family yearly income level*, *family yearly consumption level*, *occupation type* and *education level* from their home location. To the best of our knowledge, this is the first work focusing on SEA inference through the home location. The main contributions are summarized as follows:

- We enrich people's home location with knowledge from various aspects such as area-level economic statistics, housing price, point of interest (POI), and administrative division. We design multiple SEA-related features according to this knowledge. The source data of these features are mined from multiple commercial real-estate websites, official statistic bureau websites, online maps, etc.

- We propose a factorization-machine-based multi-task learning method with attention mechanism, to learn a shared representation from input features as well as attribute-specific representations for different SEA predication tasks. The multi-task method can additionally leverage the potential relationship between income, education and occupation. Comparing with existing multi-task learning methods for attribute inference, the proposed model further improves the performance with limited features by modeling the second-order feature interactions with factorization machine (FM).

- As a case study, we carry out a survey to collect people's SEAs in China. In the end, we collect a dataset that includes 9 provinces and 85 cities in China. The experiments on this dataset demonstrate that 1) home location can clearly improve the performance of predicting people's SEAs; 2) the proposed method outperforms compared methods on all SEA prediction tasks in terms of multiple metrics such as AUC and F1-measure.

The rest of this chapter is structured as follows. Section 4.2 introduces the ground-truth dataset. Section 4.3 discusses how to design and mine data for SEA-related features. The H2SEA model is proposed in Section 4.4. Experimental results are presented in Section 4.5. Section 4.6 further analyzes the relationship between housing price and income in China. The chapter is concluded in Section 4.7 with a brief discussion of limitations and directions to future research.

4.2 Ground Truth Dataset

We collected a dataset covering a sampled population's personal SEAs in 2018 in China. Each record consists of an anonymous volunteer's age, gender, home location, SEAs, etc. SEAs include **personal monthly income level, family yearly income level, personal education level and personal occupation type**. We also collected the volunteers' check-in data on a famous online social network platform called QQ. Inspired by [25], we combine the most visited check-in location during the night and collected home location to calculate the latitude and longitude of a person's home. Among 32,443 volunteers, 4,509 of them reported at least one socioeconomic attribute and agreed to share their home location for research purposes. The dataset covers 9 provinces and 85 cities in China.

Table 4.1: Demographics description

Demographics	Fraction
Personal Monthly Income	
Under 2000¥	23.04%
2000-4000¥	30.74%
Over 4000¥	20.16%
Not answer	26.06%
Family Yearly Income	
Under 40,000¥	37.19%
40,000-75,000¥	35.82%
Over 75,000¥	25.86%
Not answer	1.13%
Personal Occupation Type	
Farmers, temporary worker, unemployed, etc	62.74%
Ordinary employers, freelancer, etc	30.03%
Middle and senior managers,etc	9.63%
Not answer	0.78%
Personal Education level	
Lower secondary education	54.29%
Upper secondary education(High)	23.20%
University	22.51%

The demographics description for them is shown in Table 4.1. To protect personal privacy, we ask most volunteers to choose general socioeconomic levels. Besides, about 2,800 people also agreed to submit their exact monthly income number. In this work, we only use the accurate income ranges to calculate the correlation coefficients between income and housing price in Section 4.6. The recorded ID is a random number, which has no relationship with the volunteers' identification information. All data were collected under confidentiality agreement and only allowed for research purpose. Our dataset is temporarily not open due to the confidentiality agreement.

4.3 Feature Engineering

Given a person' features, we aim to estimate his/her *personal monthly income level*, *family yearly income level*, *family yearly consumption level*, *occupation type* and *education level*. In this work, all these problems are defined as three-level classification tasks. The latitude and longitude of the home location are too limited for multiple SEA predication. Hence, we need to design SEA-related features to enrich the home location. In this section, we introduce how to design SEA-related features as well as collecting corresponding data for these features.

4.3.1 Features based on Housing Price

A common observation is that personal income or occupation may be related to people's housing price [105, 25]. The government usually only publishes area-level average housing prices (e.g., city-level or county-level in China), which may be too course-grained for personal SEA predication. Thus, it is hard to get the exact housing price of the house which the targeted user lives in. Fortunately, some commercial real estate websites may publish the housing price of a house in or near a specific Global Positioning System (GPS) location which is now for selling. In this work, we collect the housing prices, which are near one home location, from some real estate commercial websites.

The housing price dataset is mainly crawled from Lianjia.com, which records the house prices and location information of apartments selling in China. From Lianjia, we can crawl the prices of the houses which are less than 2 kilometers away from one home location. We can find housing price information for 43% volunteers. Lianjia.com only records the prices of houses which are sold in recent time. So there may be no housing price data for one home location if no nearby houses are for selling in recent time. For missing F_{hp}, we use the nearest known housing price as a substitute if the distance is less than 10 kilometers. If there is no housing price data nearby, we use the city-level average housing price published by local governments as a substitute.

4.3.2 Features based on Renting Price

Though in this work, we mainly focus on the people who have their own house/apartment, the renting prices of an area could be also helpful in predicting people's SEAs for the following reasons. First, the number of crawled housing prices in many communities is not enough. It can not completely cover all home locations. And we observe that renting prices are usually high in high housing price areas. So renting price is an important supplement to housing prices. Second, renting prices may be related to the income or consumption level of people who have their own houses. For example, some people could gain more income by renting their house to others. It would be better to introduce more related features to alleviate the limitation of input features.

We use a similar method like housing price to collect renting prices for each home location. We also collect the renting prices from commercial websites like Lianjia.com. We can find renting price data inside the 2 km radius of 32% home location. The others are using the nearest known renting prices as an approximation.

4.3.3 Features based on Official Area-Level Economic Statistics

Features based on area-level economic statistics include several kinds of features, such as average income, Gross Domestic Product (GDP), government budget and tax. These area-level economic statistics mainly reflect the economic development level of one administrative division. Some statistics are directly related to the SEAs of people living in the area. They are usually published by governments and could be found on government websites like [17]. In some developed countries, there may be fine-grained statistics. For example, in [1], French governments publish a composite index called SEL. The SEL of a district is calculated based on the income, assets and education of people who lives in this district. The area of one district is only 1-4 km^2. However, in developing countries like China, most local governments only publish coarse-grained statistics. We find that Chinese governments only publish county-level average income for most areas. A county in China can cover hundreds of thousands of people and hundreds of square kilometers [99]. Though quite coarse-grained, these statistics could still be helpful in prediction because they are all related to the economic levels of an area in which home location belongs to. In this work, we mainly collect three types of Chinese area-level economic statistics features.

County-level average income. The published county-level average income in China covers a very large area. So for home locations in one county, their county-level average income is the same as each other.

Town-level budget and tax. The town is an administrative division smaller than the county and larger than the community or village. The town is the smallest administrative division, of which the economic statistics can be found on Chinese government websites. We cannot find town-level statistics that are directly related to personal income or occupation, like average income. In this work, we use town-level budget and tax, which may be indirectly related with people's SEAs.

4.3.4 Features based on Point of Interests

The urbanization process leads to different functional regions in a city, e.g., entertainment areas, business districts, and residential areas [107]. The function of living areas may be related to people's occupation and education level. Point of Interest (POI) can be used to give a description of the function of one area. If there are many restaurants and few schools/universities in a living area, we may think this is an entertainment area. As the number of restaurants in most areas is typically much larger than schools/universities, we need to carefully check the overall distribution. If compared to the overall distribution, the percentage of schools/universities is higher while that of the restaurants is lower, then this should be an education area rather than an entertainment area.

First, the POI information of all home locations should be collected. Then for j-th POI category, its overall frequency of all home locations is

$$OF_j = \frac{Number\ of\ POIs\ of\ the\ j-th\ category}{Number\ of\ all\ POIs}, \qquad (4.1)$$

Then the frequency of the j-th POI category in one home location H_i is calculated as:

$$of_j = \frac{Number\ of\ the\ POIs\ of\ the\ j-th\ category\ in\ H_i}{Number\ of\ all\ POIs\ in\ H_i}, \qquad (4.2)$$

Then features based on Point of Interests, F_{poi}, are:

$$F_{poi\_dis} = \{of_1/OF_1, of_2/OF_2, \ldots, of_l/OF_l\}, \qquad (4.3)$$

where l is the number of collected POI category. For a home location, if of_j/OF_j is larger than 1, it means this area has more POIs in the j-th category compared with overall distribution. Then the j-th category is more important to determine the function of this area. **POI.**

POI dataset is also crawled based on Baidu Map API Service. We collect all POI records which are less than 2 kilometers away from the home location. We can find POI information for all home locations in our dataset.

4.3.5 Categorical Features

Here we introduce several categorical features. The categorical feature is different from the above continuous features. It usually contains a number of categories or distinct groups. And there might not be a logical order between different categories or groups.

Zip code, F_{zp} can be used as a home-based feature. The zip code of a town to which a home location belongs can be found on websites like [23].

Location name can also be used in prediction. In China, we can use the **province name** (F_{pn}), **city name** (F_{cn}), **county name** (F_{con}), **town name** (F_{tn}) and **street name** (F_{sn}). They are corresponding to people's living area. Location names are useful because the gap between different places in China is quite big.

During our data collection stage, we also collect the urban type of a home location. Urban types include 3 categories: city center, city border and rural area. There are very serious urban-rural income gap and inequality in China [83]. People in the city-center may have higher income, better education and more working career opportunities than rural areas.

4.4 Home to SEA (H2SEA)

The overall architecture of the proposed method H2SEA is presented in Figure 5.1. In this section, we present the details of H2SEA. H2SEA model predicts a person's N kinds of socioeconomic attributes (denoted as $Y = \{Y^1, Y^2, \ldots, Y^N\}$) based on his/her home location (denoted as H).

4.4.1 FM-based Shared Embedding Layers

FM-based Shared Embedding Layers consists of an embedding layer and an FM layer. The features are fed into the embedding layer to get an initial representation that is shared for all tasks. Previous works usually use one feedforward neural network layer to get the initial embeddings for the input basic features. However, there is one problem: the features based on a single home location maybe not enough for predicting people's SEA. To tackle these problems, we leverage Factorization Machine (FM) [76] to generate the embeddings. Compared with the feedforward neural network layer, FM additionally considers the value of feature interactions. Feature interactions can improve SEA predication by modeling the underlying relationship between different features. Simply put,

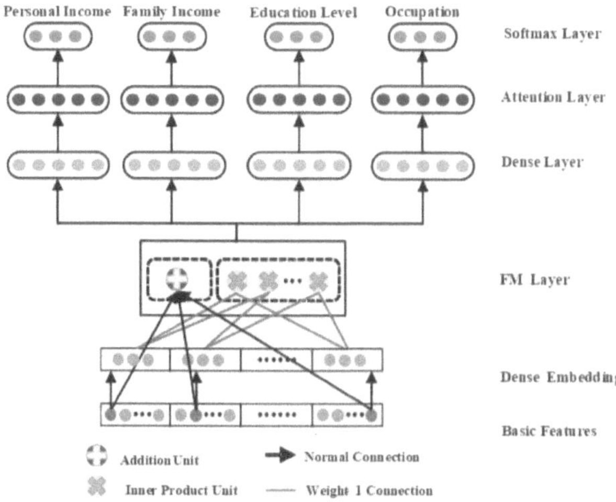

Figure 4.1: The architecture of H2SEA Model

it generates new second-order features based on basic input features. FM can automatically learn feature interactions. It embeds features into a latent space and models the second-order interactions between features via the inner product of their embedding vectors.

In this work, we denote one socioeconomic attribute of a person u as Y, the basic features as X. X includes both continuous fields (e.g., features based on housing price) and categorical fields (e.g., features based on zip code). We represent every categorical feature as a vector of one-hot encoding and every continuous feature as the value itself. So X, Y can be converted to (x, Y) where $x = [x_1, x_2, \ldots, x_i, \ldots, x_n]$. x_i is the vector representation of the i-th feature, like F_{hp}, F_{poi}, or F_{zp}. n is the number of all basic features.

FM could be seen as a combination of embedding layer and inner product layer. Here we actually use the latent feature vectors in FM as embedding network weights. The output of FM layer is the summation of an addition unit and a number of inner product units. The fm-based shared embedding e_{fm} is defined as:

$$e_{fm} = \langle w_{fm}, x \rangle + \sum_{i=1} \sum_{j=i+1} \langle V_i, V_j \rangle x_i \cdot x_j, \qquad (4.4)$$

where $w_{fm} \in R^k$ and $V_i \in R^d$. k is the dimension of one-hot vector and d is the dimension size of embedding layers. For a person, w is used to weigh its basic features' order-1 importance ((w_{fm}, x)). The latent vector V_i can measure the impact of interactions between the feature x_i and all the other features by the inner Product units. FM can train latent vector V_i (or V_j) whenever i (or j) appears in a data record.

4.4.2 Attention-based Attribute Specific Layers

Attention-based Attribute Specific Layers consist of a dense layer and an attention layer. The shared representations have captured the global signal shared by all attribute predication tasks. Next, we need to refine the shared representations to adapt to the different tasks. For each task, we use an dense layer and an attention layer to generate attribute-specific representations. First, we use the dense layer to learn an primary attribute-specific representation for the n-th SEA:

$$d^n = relu(e_{fm} \times w_d^n + b_d^n) \qquad (4.5)$$

where $relu$ is a non-linear activation. $w_d^n and b_d^n$ are weight and bias parameters for the n-th task. It is reasonable to assume that some input feature maybe more related with certain SEAs than others. For example, area-level average income may be more related with income level while the POI distribution may be more related with occupation types. To model the varying importance of features for different attribute, here we use an an attention layer:

$$t^n = relu(d^n \times w_t^n + b_t^n)) \qquad (4.6)$$

$$a^n = softmax(t^n) \tag{4.7}$$

where a^n denotes the attention weights for the n-th tasks. The sum of a^n equals to 1. The distribution of a^n can be seen as the importance of each feature embedding for the n-th transaction. The final representation for the n-th SEA predication task is the weighted sum of all shared embeddings:

$$u^n = \sum_{i=1}^{k} a_i^m \times d_i^m \tag{4.8}$$

4.4.3 Predication Layers

Predication Layers consist of all prediction layers for 4 SEA inference tasks. In the end, the output of Attention-based Attribute Specific Layer u^n is fed into the softmax (or sigmoid layer) to estimate the SEAs of a person. Take n-th SEA as an example, the predication probability $\hat{\mathbf{y}}_n$ is defined as:

$$\hat{y}^n = softmax(u^n) \tag{4.9}$$

If the distribution of attributes is even, the loss function $\mathcal{L}_n$ is computed as follows:

$$\mathcal{L}_n = -\frac{1}{M_n} \sum_{j=1}^{M_n} \sum_{k=1}^{C_n} y_{j,k}^n \log(\hat{y}_{j,k}^n) \tag{4.10}$$

where M_n is the number of users whose n-th SEA is not missing. C_n is the number of n-th attribute category. $y_{j,k}^n$ and $\hat{y}_{j,k}^n$ are the ground truth and estimated SEA labels, respectively.

If the distribution of a SEA is quite imbalanced, we leverage a weighted cross-entropy function to calculate the prediction loss $\mathcal{L}_n$ as follows:

$$\mathcal{L}_n = -\frac{1}{M_n} \sum_{j=1}^{M_n} \sum_{k=1}^{C_n} w_{y_{j,k}^n} y_{j,k}^n \log(\hat{y}_{j,k}^n) \tag{4.11}$$

where $w_{y_{j,k}^n} = \frac{\sum_{k=1}^{C_n} \sqrt{M_n^k}}{\sqrt{M_n}}$ is a parameter to control the cost weight of each attribute category, M_n^k is the number of people with the n-th attribute label. The total loss of all SEA tasks can be defined as:

$$\mathcal{L}_{\text{total}} = \sum_{n=1}^{N} \lambda_n \mathcal{L}_n + \alpha||\Theta|| \tag{4.12}$$

where λ_n is are hyper-parameters controlling the relative importance of the n-th SEA predication task. We enforce that $\sum_{n=1}^{N} \lambda_n = 1$ to facilitate the tuning of the hyper-parameters. Θ denotes all trainable parameters of H2SEA model. We adopt L2-normalization [100] and dropout [89] to prevent overfitting. α controls the L_2 regularization strength. By optimizing the entire loss $\mathcal{L}_{\text{total}}$, our model can get the best results for recommending task. H2SEA model is trained via back-propagation and Adam [55].

4.5 EXPERIMENTS

In this section, through experiments based on the actual dataset, we want to answer the following questions: 1) Whether home location has predictive power for socioeconomic attributes? 2) Whether H2SEA model outperforms widely-used baselines? 3) What are the most important home-based features for income, occupation or education predication? 4) How different settings (e.g. dropout, λ_n) affect the performance?

4.5.1 Experiment Setup

We use the following SEA prediction tasks to test the predictive power of home location.

Personal Income Level (PIL). Three-level personal income prediction task. The boundary lines are 2,000 yuan and 4000 yuan every month.

Table 4.2: Performance Comparison

Task	Method	F1	Auc	G-Mean	Acc
Personal Income	POP	0.3329	0.4894	0.3521	0.3561
	LR	0.5314	0.7231	0.4838	0.5233
	Xgboost	0.5734	0.7482	0.4652	0.5296
	ETNA	0.5856	0.7630	0.4719	0.5394
	H2SEA	**0.5999**	**0.7786**	**0.5020**	**0.5501**
Family Income	POP	0.3247	0.4978	0.3729	0.3648
	LR	0.4815	0.7035	0.4354	0.5233
	Xgboost	0.5050	0.7181	0.4676	0.5296
	ETNA	0.5183	0.7351	0.4741	0.5434
	H2SEA	**0.5345**	**0.7546**	**0.5261**	**0.5576**
Education Level	POP	0.3259	0.4997	0.5529	0.5463
	LR	0.4825	0.7449	0.5676	0.6006
	Xgboost	0.4927	0.7697	0.6568	0.6585
	ETNA	0.5083	0.7975	0.6645	0.6595
	H2SEA	**0.5272**	**0.8289**	**0.7039**	**0.6640**
Occupation Type	POP	0.3391	0.5075	0.5562	0.5881
	LR	0.4681	0.7088	0.5618	0.5858
	Xgboost	0.4717	0.6997	0.5835	**0.5952**
	ETNA	0.4848	0.7201	0.5877	0.5833
	H2SEA	**0.5003**	**0.7434**	**0.6633**	0.5869

The percentage of people in low-income-level is 31.2%, middle-income-level is 41.5 % while high-income-level is 27.3%.

Family Income Level (FIL). Three-level family income prediction task. The boundary line are 40,000 ¥and 75,000 ¥every year. **Education Level (EL).** Three-level education level prediction task. This task aims to predict whether a person has a university degree, high school degree or junior high school degree. The percentage of junior high school is 54.29%, high school degree is 23.20% while university is 22.51%.

Occupation Type (OT) Three-level occupation prediction task. This task aims to predict people's occupation types. The people in low-level (farmers, temporary worker, unemployed) is 62.74%, middle-level (ordinary employers, the freelancer) is 30.03% while high-level(manager) is 9.63%.

Evaluation Metrics. We use the following evaluation metrics: macro-F1, AUC, G-mean and Accuracy for all tasks. In unbalanced tasks like **OT**, macro-F1 is the most important metric.

Baselines. To the best of our knowledge, there exists no model focusing on estimating personal SEAs from home location. Here we use the following widely-used standard classification methods as baselines:

Popular (POP): POP simply estimate an individual' SEA as the majority classes [96]. This model ignores all input features.

LR: We use 2-degree Logistic regression (LR) to model the linear combination of basic features and all order-2 feature interactions.

Gradient boosting decision tree (GBDT). The gradient boosting model is famous for its outstanding performance and efficiency for general classification tasks. Xgboost is an open-source gradient boosting library [20]. We use all features to train Xgboost model.

Embedding Transformation Network with Attention (ETNA). This is the state-of-the-art multi-task demographic model. It also uses the attention mechanism to refine the shared embeddings for different demographics. However, compared with H2SEA, ETNA neglects the effect of feature interactions.

70% of people are chosen as training dataset, 20% as validation dataset and 10% as test dataset. Our model is implemented based on Keras [22]. Hyper-parameters of H2SEA are tuned by grid-searching on the validation set. Due to limited space, here we only show the best settings of **PIL** as an example. The latent dimension of FM component (or field embedding size) is 6. The dropout is 0.3, the number of neurons per layer (deep component) is 32, the number of hidden layers (deep component) is 3. The learning rate of Adam is 0.001, the activation function is relu, and L2-norm ratio is 0.00001. Our model is implemented with Keras [22], and trained on a 64-bit server with 2 NVIDIA 1080Ti GPU.

4.5.2 Results Analysis

This section mainly answers whether personal SEAs can be predicted based on home location and how H2SEA model performs compared with baselines.

The results of all tasks are showed in Table 4.2. The numbers in the Table 4.2 are averaged by 10 times of train-testing. To achieve the best performance, we conducted carefully parameter tuning of all methods, which is introduced in Section 5.3. From Table 4.2, we have following observations.

Home location clearly improves the performance in estimating Personal Income, Family Income, Occupation and Education level. Especially, compared with Random Guess, H2SEA model can increase 80.22% in F1-score and 42.57% in G-Mean in personal income prediction; 64.57% in F1-score and 41.08% in G-Mean in family income prediction; 61.76% in F1-score and 27.31% in G-Mean in education level prediction; 47.55% in F1-score and 19.26% in G-Mean in occupation prediction.

Considering the relative improvements compared with Random Guess, personal income level achieves the best results than family income, occupation type and education. It is quite surprising that home location achieves weaker results in family income than personal income. Because a house/apartment is often bought by a family rather than one individual, the housing price may be more related to the family income level than the personal income level. We conjecture the weak predictability may be caused by the weak relationship between housing price and family/personal income level. The most important feature for income is county-level average personal income, which is clearly more related to personal income level. We will further analyze why the

relationship between housing prices and income is weak in Section 6.
However, we should note that the H2SEA model still performs much
better than random guess. The performances of occupation type and
education level are weaker than income prediction, indicating that home
location alone is not enough to predict these two attributes. Besides,
the imbalance of these two attributes also increases the difficulty in
estimation.

**H2SEA model outperforms all baselines in terms of F1-score and
G-Mean**. The second best classifier is ETNA. H2SEA outperforms
ETNA in all tasks by 2.43% - 3.71%, 2.70% - 4.24%, 6.06% - 12.94%
and 0.62% - 2.84% in terms of F1-score, AUC, G-Mean and accuracy, respectively. It indicates that second-order feature interactions can clearly
improve performance. ETNA is better than all the other single-tasks
models, like Xgboost and Logistic Regression (LR). It demonstrates that
the multi-task learning method can model the underlying relationships
between various attributes. It is worth to point out that the accuracy of
H2SEA is worse than Xgboost in Occupation level by 1.39%. This is
caused by imbalance. Only 9.63 percent of people are in higher level
(Middle and senior managers, etc). We mainly consider more about
AUC, macro-F1 and G-Mean in an unbalanced task. For example, the
G-Mean of H2SEA is 13.67% better than Xgboost. Besides, the G-Mean
of H2SEA is 12.94% better than ETNA in occupation estimation compared to only 6.06% in personal income level estimation. This indicates
that H2SEA may better handle imbalanced datasets through weighted
softmax loss function.

4.5.3 Feature Importance Analysis

This section discusses the most important features of each task. We
mainly show the metrics of the top 5 important features in each task.
The metrics are calculated when only using one feature for prediction.
The importance of home-based features can help to understand the relationship between home location and different SEAs. In Table 4.3, for

Table 4.3: The Metrics of Top 5 Features in Each Task

Task	Feature	F1	Auc	G-Mean	Acc
Personal Income	county-income	0.4752	0.7434	0.4131	0.4576
	POI	0.4767	0.7468	0.4149	0.4566
	city-name	0.4787	0.7361	0.4161	0.4171
	province-name	0.4338	0.7295	0.3794	0.3984
	Average Housing Price	0.3623	0.6915	0.3653	0.3331
Family Income	county-income	0.4609	0.6833	0.4750	0.5031
	POI	0.4445	0.6572	0.4713	0.4930
	city-name	0.4214	0.6393	0.4706	0.4774
	province-name	0.4237	0.6727	0.4557	0.4376
	county-name	0.3847	0.6082	0.4582	0.3961
Education level	POI	0.5061	0.6647	0.6686	0.5255
	urban type	0.5224	0.5964	0.7018	0.5058
	county-income	0.4839	0.6441	0.6280	0.4910
	county-name	0.5425	0.4952	0.6863	0.4417
	Average Housing Price	0.5296	0.4736	0.6924	0.4271
Occupation Type	POI	0.4641	0.6839	0.6073	0.5097
	county-income	0.4751	0.6459	0.6324	0.5411
	city-name	0.4803	0.5877	0.6511	0.4796
	urban type	0.4346	0.6454	0.5771	0.4781
	county-name	0.5057	0.5409	0.6085	0.4344

each task, the importance of features is decreasing from top to bottom. The importance is mainly ordered by the combined improvement of F1, AUC, and Gmean. From Table 4.3, we can get following observations.

County-income is the most important feature of income prediction. It shows that even coarse-grained area-level income statistics may be of great help for income prediction. Besides, county-income is also the second important feature for occupation prediction and the third important feature for education prediction. This result indicates that the county-income is highly related to people's occupation and education level. This is reasonable because some occupation types earn much more money than others, and the education resources in high-income-level-areas are usually richer than low-income-level-areas.

POI is the most important feature for education and occupation prediction. It is also the second important feature of income prediction. POI reflects the function of the living areas. The results demonstrate that the function of one living area is highly related to people's occupation and educational background. For example, we find that people with university degrees are more likely to lives in the areas, where the most important POI categories are related to universities, governments or high-tech companies.

Housing prices are not so effective in SEA prediction tasks. Housing prices may be one of the most widely used home-based features and are often used as a proxy of people's income in previous works like [49]. This is mainly because people usually believe that housing price is highly related to income. However, our study shows the average housing price is only the fifth important feature for personal income and education level prediction. This may be caused by data missing. Besides the data missing problem, we also analyze other possible reasons in Section 4.6.

4.6 Relationship between Housing Price and Income

Researchers usually think that housing price is a very important feature when studying the relationship between home and socioeconomic attributes. Richer people live in high price-level areas and poorer people live in low-price-level areas. Previous studies like [105] also show that the housing price has a strong correlation with personal income in Singapore. However, in our case, housing price is not so effective in prediction. Here we try to give an analysis of possible reasons.

The first possible reason is that China is still a developing country. Different from Singapore (city-level), China includes many undeveloped cities. Based on our dataset, the correlation coefficient between housing prices and personal income over China is only 0.185, much weaker than Singapore (0.8 [105]). So how does the level of development affect the relationship?

The second reason is caused by the limitation of data collection. China's house market is still not very mature. We find an interesting phenomenon: **many low-income people seem to live in high-price areas**. After communicating with corresponding researchers, we learn that there are generally 2 kinds of houses/apartments in China. We can only find the housing prices of one kind on websites. These houses/apartments are "commercial houses", which are the same as those in developed countries. They are usually built by real estate companies and bought by new middle classes. Their prices can be easily found on legal commercial websites. And the prices are highly related to customers' income levels. The other one is built by the farmers themselves. These houses are not free for buying and selling like the first kind and their prices are much lower. The prices of the second kind are hard to find on open websites. Because we cannot find the actual prices of the second kinds, it turns out that many low-income people seem to live in high-priced areas.

4.7 Chapter Summary

This chapter focuses on examining whether people's multiple socioeconomic attributes (e.g, income and occupation) can be estimated only based on their home location. This study first designs and collects multiple types of SEA-related features such as housing price, county-level income and urban types. Then an FM-based multi-task learning method named H2SEA is proposed to model both second-level feature interactions to further improve the prediction accuracy. Based on a dataset collected in 9 provinces of China, the experiment shows that home location and home-based features can clearly improve the performance in predicting people's income, education and occupation. And H2SEA model outperforms the compared methods in terms of various metrics like AUC and F1.

Chapter 5

Multi-Task-level Problem: Improving User-Attribute-Enhanced tasks by Attribute Inference

In this chapter, we mainly discuss a Multi-Task-Level problem including UAI and its downstream task (i.e., UAE): improve the performance of UAE by UAI. As a case study, we focus on improving the performance of CF recommendation with missing attributes by auxiliary UAI tasks.

Contents

5.1 Introduction

Nowadays, there are tremendous music, products and movies for users to pick. Recommender system is a crucial tool to provide personalized recommendation services for users to tackle "information overload" problem [78]. Among recommender systems [79, 27, 81], collaborative filtering (CF) is one of the most successful techniques. It assumes a user would tend to show similar preference on items which are liked by other similar users. Recently, inspired by the recent success of Graph Convolutional Network (GCN) on graph standard GCN[56, 106], a couple of GCN-based CF algorithms have been proposed [11, 106, 98, 45]. For example, GC-MC [11] applies GCN on user-item graph to exploit the direct connections between users and items. NGCF [98] improves the recommendation performance by modeling high-order connectivity on a user-item graph. More recently, LightGCN achieves state-of-the-art performance by simplifying feature transformation and nonlinear activation in GCN layers [45].

Though CF methods achieve great success in a wide range of scenarios, sometimes they may encounter *interaction sparsity problem*. In realistic recommendation scenarios, many users often only interact with a very small proportion of items. The few interactions of these users are insufficient for CF to learn their accurate preference on items. To alleviate the problem, various attributes of user (e.g., gender, age, location) and item (e.g., category, genres, brands) have been exploited to improve the original CF methods [82, 57]. For CF methods only relies on the user-item interaction data, we refer them as pure CF methods [97]. For CF methods also leveraging attributes, we refer them as attribute-enhanced CF methods. Though several GCN-based methods such as GCMC [11] leverages attributes to enhance recommendation, most GCN-based CF methods including LightGCN and NGCF are pure CF methods [98, 45] until now. Moreover, there is still one problem remaining for attribute-enhanced methods: missing user/item attributes.

Table 5.1: Impact on Recommending Performance with Increasing Missing Rates

Missing	Yelp-OH		Yelp-NC		Tianchi	
	recall	relative-decrease	recall	relative-decrease	recall	relative-decrease
0%	0.0902	0.0%	0.0795	0.0%	0.0285	0.0%
10%	0.0902	0.0%	0.0772	-2.9%	0.0275	-3.5%
20%	0.0877	-2.8%	0.0772	-2.9%	0.0256	-10.2%
30%	0.0855	-5.2%	0.0770	-3.1%	0.024	-15.8%
40%	0.0862	-4.4%	0.0723	-9.1%	0.0227	-20.4%
50%	0.0831	-7.9%	0.0746	-6.2%	0.0203	-28.8%
60%	0.0814	-9.8%	0.0726	-8.7%	0.0204	-28.4%
70%	0.0807	-10.5%	0.0724	-8.9%	0.0192	-32.6%
80%	0.0833	-7.7%	0.0701	-11.8%	0.0191	-33.0%
90%	0.0803	-11.0%	0.0693	-12.8%	0.0187	-34.4%
99%	0.0785	-12.9%	0.0661	-16.9%	0.0175	-38.6%

Attributes are often missing in real-world scenarios. For instance, many users are reluctant to provide age or location information due to privacy concerns. Researchers often use zeros, average values or special tags as substitutes for missing values. This method can make attribute-enhanced CF methods easily adaptive to incomplete attribute features. However, their performance may be affected if the missing rate is too high. Here we quantify the negative impact caused by missing attributes through experiments. The experiments shows how the performance of NFM [75] is affected, when the missing rates of Yelp-OH, Yelp-NC and Tianchi datasets increases from 0% to 90%[1]. Table 5.1 shows the performance (in terms of recall@20) are more and more seriously affected with increasing missing rates. Consequently, this work aims to: **1) enhance GCN-based pure CF methods with attributes, and 2) reduce the negative impact caused by missing attributes**.

User/item profiling, which aims to estimate the attribute of user/item, is another important task for online platforms [3]. Profiling and recommending are usually two separate tasks. Here we argue that combining these two tasks into one multi-task learning [18] framework may potentially improve the performance of recommender systems with missing attributes. First, user/item attributes can be predicted based on user-item

---

[1] The details on the datasets and metrics are described in Section IV.

 Chapter 5 Multi-Task-level Problem: Improving User-Attribute-Enhanced tasks by

interactions, which is also the source data for recommending. Second, from the perspective of GCN, the two tasks are both graph node representation learning tasks by modeling node interactions. Third, the estimation from user/item profiling task is usually more accurate than simple substitutes. Therefore, it is plausible to alleviate the missing attributes problem for recommendation by taking user/item profiling as an auxiliary task.

In this chapter, we first verify whether combining user/item profiling together into recommending can alleviate the missing attributes problem for GCN-based CF models. Based on this exploration, we develop a new Attribute-enhanced GCN (AEGCN) method, and take a recent GCN-based pure CF model, LightGCN as our base model. We define both recommending and profiling tasks in one user-item bipartite graph. Firstly, ID and attributes features of users and items are represented as node embeddings through an feature embedding layer. Then we leverage the graph convolution (GC) layers of the base model to learn the user/item-hidden representations. The GC layers perform graph convolution operations on the user-item interactions to refine the user/item node embeddings. At last the embeddings learned at the feature embedding layer and GC layers are combined to obtain the final representation for recommending task. Simultaneously, the final user and item representation are fed into dense layers to estimate users' and item's missing profiles. Our main contributions can be summarized as follows:

- We highlight the missing attributes problem by quantifying the negative impact of the missing attributes on the recommending performance through empirical studies.

- We propose AEGCN, an end-to-end multi-task GCN-based CF method, which improves recommending performance with incomplete attributes by auxiliary user/item profiling tasks.

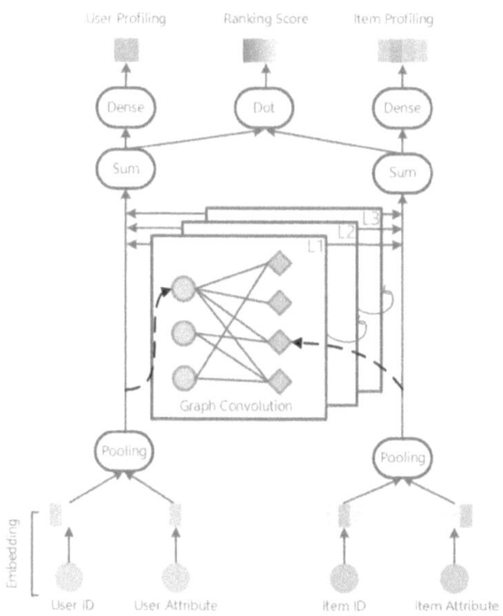

Figure 5.1: An illustration of Model architecture.

- We conduct extensive experiments on three real-world datasets, which demonstrate the effectiveness of AEGCN in alleviating the missing attributes problem.

The rest of this chapter is organized as follows. Section 5.2 describes AEGCN model in details. The efficiency of AEGCN model is demonstrated with experimental evaluation in Section 5.3. Finally, Section 5.4 concludes the Chapter 5.

5.2 Methodology

The overall architecture of the proposed method is presented in Figure 5.1. Simply put, AEGCN jointly learns to predict the preference ranking score of a target user for an item, and to estimate the attribute for a user or item. In this section, we present the details of AEGCN.

5.2.1 Feature Embedding Layer

The feature embedding layer aims to form initial low dimensional embeddings of user (or item) from their ID and attribute features. For each ID and attribute feature, we associate it with an embedding vector with same length. If one attribute is missing, we use zeros vector as its initial embedding. Then we pool the initial embeddings of one user (or item) into one vector. The vector of each user or item will be fed into the following graph convolution layers. We tried different pooling methods (e.g, sum pooling and attention pooling). There is no clear difference to recommending performance, especially when the missing rate of attribute is high. Here average pooling is adopted. Assume u denotes a user and i denotes an item. $|u|$ denotes the number of nonzero features in u, and $\mathbf{E}_U \in \mathbf{R}^{U \times T}$ is the embedding matrix for users. U denotes the number of user features and T denotes the embedding size. $\mathbf{e}_u^{(0)}$ denotes the initial pooling vector for u. Similarly, $\mathbf{E}_I \in \mathbf{R}^{I \times T}$ is the embedding matrix for item features. The initial pooling vector for item i is $\mathbf{e}_i^{(0)}$. $\mathbf{e}_i^{(0)}$ and $\mathbf{e}_i^{(0)}$ are defined as:

$$
\begin{aligned}
\mathbf{e}_u^{(0)} &= \tfrac{1}{|u|}\mathbf{E_u}^T u, \\
\mathbf{e}_i^{(0)} &= \tfrac{1}{|i|}\mathbf{E_i}^T i.
\end{aligned}
\tag{5.1}
$$

5.2.2 Graph Convolution Layers

The graph convolution (GC) layers aim to refine the initial user/item embeddings by modeling high-order connectivity relations. Our main motivation is to investigate whether the GCN-based pure CF methods can be improved by the auxiliary profiling task. Here we leverage the graph convolution layers of LightGCN [45] to generate user/item graph embeddings. In graph convolution layers, initial user/item embeddings are enhanced by propagated through the user-item interaction graph. This can augment the user and item representations with explicit collaborative filtering signal [98, 45]. The graph convolution operation is defined as:

$$\begin{aligned}
\mathbf{e}_u^{(k+1)} &= \Sigma_{i \in \mathcal{N}_u} \frac{1}{\sqrt{|\mathcal{N}_u|}\sqrt{|\mathcal{N}_i|}} \mathbf{e}_i^{(k)}, \\
\mathbf{e}_i^{(k+1)} &= \Sigma_{u \in N_i} \frac{1}{\sqrt{|\mathcal{N}_i|}\sqrt{|\mathcal{N}_u|}} \mathbf{e}_u^{(k)},
\end{aligned} \tag{5.2}$$

where $\mathbf{e}_u^{(k)}$ denotes the user u's graph embedding on the $k-th$ GC layer, $|\mathcal{N}_u|$ denotes the number of items that u interacts, and $|\mathcal{N}_i|$ denotes the number of users interacts with i. The symmetric normalization term $\frac{1}{\sqrt{|\mathcal{N}_u|}\sqrt{|\mathcal{N}_i|}}$ can avoid the scale of graph embeddings increasing with graph convolution operations [56].

5.2.3 Recommending Layer

The recommending layer aims to collect user/item graph embeddings generated from at each layer and predicts the ranking score of a user-item pair. After K layers, we sum up the initial embeddings and the graph embeddings of each layer to form the final representation of a user (or an item):

$$\begin{aligned}
\mathbf{e}_u &= \alpha_0 \mathbf{e}_u^{(0)} + \alpha_1 \mathbf{e}_u^{(1)} + \ldots + \alpha_K \mathbf{e}_u^{(K)}, \\
\mathbf{e}_i &= \alpha_0 \mathbf{e}_i^{(0)} + \alpha_1 \mathbf{e}_i^{(1)} + \ldots + \alpha_K \mathbf{e}_i^{(K)}, \\
\text{s.t.} \quad & \alpha_k \geq 0, \quad \text{and } \alpha_k = 1/(K+1)
\end{aligned} \tag{5.3}$$

The model prediction for the ranking score of a user-item pair is defined as the inner product of user and item final representations:

$$\hat{y}_{ui} = \mathbf{e}_u^T \mathbf{e}_i \tag{5.4}$$

5.2.4 Profiling Layers

The profiling layers aim to use the final user (or item) representation to estimate the attribute of a user (or item). We use a dense layer and a softmax function to get the estimated attribute. Dropout is applied on the dense layer to prevent overfitting for profiling tasks. For example, a probability distribution over possible user attribute categories $\hat{\mathbf{y}}_u$ is defined as:

$$\hat{y}_u = \text{softmax}(\text{ReLU}(\mathbf{e}_u \times \mathbf{W}_\mathrm{u} + \mathbf{b}_\mathrm{u})) \tag{5.5}$$

where $\mathbf{W}_\mathrm{u}$ and $\mathbf{b}_\mathrm{u}$ are the weight and bias of a dense layer, respectively. The item attribute probability vector $\hat{y}_i$ can be computed similarly.

5.2.5 Model Training

To learn the model parameters for recommending tasks, we employ the Bayesian Personalized Rankin (BPR) loss [77]. BPR focuses on the relative order between observed and unobserved user-item interactions. It assigns higher ranking scores for observed instances than unobserved ones. In this work, we optimize the regularized BPR loss:

$$L_{BPR} = -\sum_{u=1}^{U} \sum_{i \in \mathcal{N}_u} \sum_{j \notin \mathcal{N}_u} \ln \sigma(\hat{y}_{ui} - \hat{y}_{uj}) + \mu \|\Theta\|^2 \tag{5.6}$$

where Θ includes all the trainable model parameters ($\mathbf{E}_U$, $\mathbf{E}_I$, $\mathbf{W}_u$, $\mathbf{W}_i$, $\mathbf{b}_u$, $\mathbf{b}$). μ controls the L_2 regularization strength to prevent overfitting. For each observed (u, i), we pair it with an unobserved item j.

If the distribution of attributes is even, the standard cross-entropy can be computed as follows. Take user attribute prediction task as an example, the loss function $\mathcal{L}_u$ we used is the standard cross-entropy, which is computed as follows:

$$\mathcal{L}_u = -\frac{1}{N_A} \sum_{j=1}^{N_A} \sum_{k=1}^{C_A} y_{j,k}^A \log(\hat{y}_{j,k}^A) \tag{5.7}$$

where N_A is the number of users whose attribute is not missing. C_A is the number of user attribute category. $y_{j,k}^A$ and $\hat{y}_{j,k}^A$ are the ground truth and estimated user attribute labels, respectively.

If the distribution of attributes is quite imbalanced, which may lead to the majority voting classification errors. Actually we found that this phenomenon occurs in all of our experiments datasets.Here, we utilize a weighted cross-entropy function as the attributes prediction loss $\mathcal{L}_u$, which is formulated as follows:

$$\mathcal{L}_u = -\frac{1}{N_A} \sum_{j=1}^{N_A} \sum_{k=1}^{C_A} w_{y_{j,k}^A} y_{j,k}^A \log(\hat{y}_{j,k}^A) \tag{5.8}$$

where N_A is the number of users with known attribute labels, C_A is the number of attribute category. $y_{j,k}^A$ and $\hat{y}_{j,k}^A$ are the ground-truth labels and estimated , respectively. $w_{y_{j,k}^A} = \frac{\sum_{k=1}^{C_A} \sqrt{N_A^k}}{\sqrt{N_A}}$ is a parameter to control the cost weight of each attribute category, N_A^k is the number of users with the k_{th} attribute label. The total loss of the entire network can be defined as:

$$\mathcal{L}_{\text{total}} = \mathcal{L}_{\text{BPR}} + \lambda_u \mathcal{L}_{\text{u}} + \lambda_i \mathcal{L}_{\text{i}} \tag{5.9}$$

where λ_u and λ_i are hyper-parameters controlling the relative importance of the auxiliary user and item profiling tasks. By optimizing the entire loss $\mathcal{L}_{\text{total}}$, our model can get the best results for recommending task.

5.3 EXPERIMENTS

We evaluate experiments on three real-world datasets, aiming to answer the following research questions:

- RQ1: Compared with state-of-the-art pure CF and attribute-enhanced CF methods, how does AEGCN perform on top-K recommendation as the missing rate of attributes increase?

Table 5.2: Dataset Description

Dataset	Yelp-OH	Yelp-NC	Tianchi
#User	23,637	29,115	101,955
#Item	14,002	14,042	79,575
#Interactions	212,132	268,917	723,977
Density	0.0641%	0.0658%	0.0089%
#User Category	151	36	-
#Item Category	195	55	301

- RQ2: How do the effectiveness of attributes affect the performance of AEGCN?

- RQ3: How do different settings (e.g., λ) affect the recommending performance of AEGCN in with missing attributes?

- RQ4: How does AEGCN perform on user/item profiling tasks?

5.3.1 Dataset Description

To demonstrate the effectiveness of AEGCN, we conduct experiments on datasets from Yelp and Tianchi. They are publicly accessible and vary in terms of domain, size, and sparsity. The three-core setting is adopted on all datasets to ensure data quality, i.e., retaining users with at least three interactions [42]. We summarize the statistics of datasets in Table 5.2.

- **Yelp-OH & Yelp-NC**. These are two subsets of the Yelp Challenge Dataset (2018)[2]. One interaction means a user visits/reviews local businesses, like bars and restaurants. These local businesses are viewed as items in our work. In particular, we extract user-item interactions in two different areas of USA – North Carolina and Ohio – to construct datasets, named Yelp-NC and Yelp-OH respectively. The attribute of an item is the city where it locates. The item attribute includes 195 and 55 categories for Yelp-OH and Yelp-NC respectively. We designed a new user attribute: the most frequent city which a user visits. It can be generated based on the items which a user visits in the training dataset. The user attribute includes 151 and 36 categories for Yelp-OH and Yelp-NC respectively.

- **Tianchi**. This is a public dataset opened by Alibaba's competition[3] "User Behavior Diversities Prediction", which is based on the real users-items behavior data on Alibaba's E-Commerce platforms. After data cleaning, the dataset includes 723,977 interactions of 101,955 customers on 79,575 items. In this dataset, one interaction means a customer purchased an item. We use the brands of items as items' attributes. The item attribute includes 301 categories, which consists of the most frequent 300 brands and "other brands".

If the distribution of one attributes is even, then the number of the attribute categories usually keeps the same even the missing rate is very high. However, the distributions of user/item attributes in all of our datasets are highly imbalanced. As a result, when the missing rate of user/item attribute is very high (e.g, more than 80%), the instances of some categories may be all missing. For example, 144 categories of item attribute in Yelp-OH are all missing when the missing rate is 99%. In Section IV, 99% missing rate means randomly 99% of all attributes in one dataset are missing. In our experiments, we reduce the number of categories for final prediction to those with known attributes in the training dataset. For Yelp-OH, there are only 51 (195-144 = 51) categories which can be predicted for the item profiling component.

To evaluate the performance of top-K recommendation, we adopted the leave-one-out evaluation, which has been widely used in literature [47, 10]. For each user, we hold-out his/her latest interaction as the testing set and utilized the remaining data for training. From the training set, we select the latest interaction as validation set to tune hyper-parameters. For each observed user-item interaction, we treat it as a positive instance,

Table 5.3: Results comparison with 0% and 99% missing attributes

Pure CF	Yelp-OH		Yelp-NC		Tianchi		Yelp-OH		Yelp-NC		Tianc
	recall	ndcg	recall	ndcg	recall	ndcg	recall	ndcg	recall	ndcg	recall
GCMC	0.0418	0.0167	0.0418	0.0167	0.0100	0.0040	0.0418	0.0167	0.0418	0.0167	0.0100
PINSAGE	0.0491	0.0193	0.0486	0.0193	0.0136	0.0055	0.0491	0.0193	0.0486	0.0193	0.0136
NGCF	0.0824	0.0330	0.0728	0.0288	0.0323	0.0143	0.0824	0.0330	0.0728	0.0288	0.0323
LightGCN	**0.0948**	**0.0391**	**0.0871**	**0.0330**	**0.0390**	**0.0172**	**0.0948**	**0.0391**	**0.0871**	**0.0330**	**0.0390**
Enhanced	Complete Attributes						99% Attributes Missing				
GCMC-P	0.0552	0.0216	0.0596	0.0233	0.0117	0.0043	0.0481	0.0193	0.0514	0.0192	0.0113
CC-CC	0.0842	0.0335	0.0714	0.0277	0.0207	0.0087	0.0704	0.0277	0.0651	0.0257	0.0172
NFM	0.0902	0.0361	0.0795	0.0316	**0.0285**	**0.0120**	**0.0785**	**0.0321**	0.0661	0.0262	**0.0175**
DIN	**0.1073**	**0.0414**	**0.0895**	**0.0355**	0.0231	0.0092	0.0769	0.0307	**0.0803**	**0.0318**	0.0159
AEGCN	**0.1099**	**0.0416**	**0.0914**	**0.0378**	**0.0432**	**0.0191**	**0.0988**	**0.0407**	**0.0892**	**0.0362**	**0.0397**

and then conduct the negative sampling strategy to pair it with one negative item that the user did not interact before. For profiling part, we keep 95% of known attributes as training set, the other 5% as testing set. For all missing attributes, we use an "unknown" category to represent them.

5.3.2 Experiments Settings

Evaluation Metrics

For each user in the testing set, we treat all the items that the user has not interacted with as negative items. Then each method outputs the user's preference scores over all the items, except the positive ones in the training set. To evaluate the effectiveness of top-K recommendation and preference ranking, we adopt two widely-used evaluation metrics [46, 43]: recall@K and ndcg@K. By default, we set $K = 20$. We report the average metrics for all users in the testing set. Recall intuitively measures whether the test item is in the top-20 list, and the ndcg accounts for the position of the hit by assigning higher scores to hits at top ranks. We calculated both metrics for each test user and reported the average score.

Baselines

We compare our method with several state-of-the-art CF methods as follows. First, we introduce four GCN-based pure CF methods which neglect attributes.

- **GCMC** [11]. This model is one of the earliest attempts to apply GCN on user-item graph. It employs one convolutional layer to exploit the direct connections between users and items.

- **Pinsage** [106]: PinSage is a famous industrial solution for large-scale graph recommending task. It is designed to employ GraphSAGE [38] on item-item graph. We apply GraphSAGE on user-item interaction graph like [98]. The GCN layer depth is set to 2 as suggested in [106].

- **NGCF** [98]. This model improves the performance by incorporating high-order connectivity in user-item interaction graph into embeddings. The depth of GC-layer is 3 as suggested in [98].

- **LightGCN** [45]. Compared with NGCF, LightGCN achieves state-of-the-art performance by simplifying feature transformation and nonlinear activation in GCN layers. The depth of GC-layer is 3 as suggested in [45]. It is also the base GCN model for AEGCN.

Next, we introduce four attribute-enhanced methods.

- **GCMC-P** [11]: GCMC [11] can also add user/item attributes into input. To distinguish from GCMC without attributes, we refer attribute-enhanced GCMC as GCMC-P (GCMC with profiles). The number of hidden units in the dense layer for attribute embedding is 128 for Yelp-OH and Tianchi, while 32 for Yelp-NC.

- **NFM** [44]: This model is proposed to use a bi-interaction layer and a multi-layer perceptron (MLP) to capture the nonlinear and high-order interaction between user/item IDs and attributes. The depth of MLP layer is set to 1 as suggested in [44].

- **CC-CC** [80]: This model leverages Random Feature Sampling and Adaptive Feature Sampling strategies to handle the missing feature values. We follow the same searching space suggested in [80]. For example, the Cold Sampling and Feature Sampling ratios are searched between 0.0 to 0.2; the coefficient of L_2 normalization is searched between 1e-5 and 1e-3.

- **DIN** [113]: This model exploits the relationship between users' historical interactions and the target item. It uses attention mechanism to learn the representation of users' historical interactions w.r.t. the target item. Beside interactions, DIN can take user/item attribute into consideration.

Parameter Settings

We implement AEGCN in Tensorflow. The user/item embedding size is fixed to 64 for all models. We optimize all models with the Adam optimizer. The mini-batch size is 2,048 for all datasets. The Xavier initializer [35] is applied to initialize the model parameters. We apply a grid search to find the best hyper-parameters. The learning rate is tuned amongst 0.1, 0.01, 0.001, 0.0001, the coefficient of L_2 regularization for all models except CC-CC is searched in $\{10^{-5}, 10^{-4}, \cdots, 10^{-1}\}$. The depth of the GC-layer in AEGCN L is three. The depth of the dense layer for user/item profiling is one for all datasets. The number of hidden units of this dense layer is 128 for Yelp-OH and Tianchi, while 32 for Yelp-NC. For AEGCN, we only apply dropout technique on the dense layer for profiling. The dropout for all models is searched between 0.0 and 0.8. λ_u, λ_i are searched in $\{1, 0.1, 0.01, 0.001\}$ when recommending is the main task. Besides, the early stopping strategy is performed like in [98]: training is stopped if recall@20 on the validation data does not increase for 50 successive epochs. Typically, 500 epochs

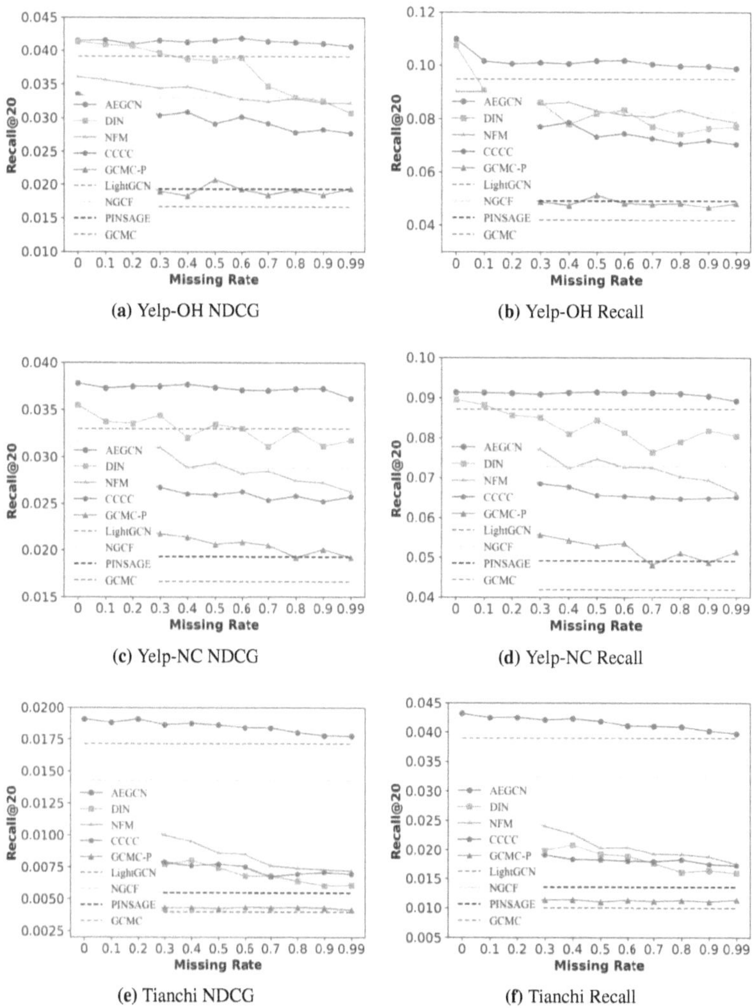

Figure 5.2: The influence of missing attributes on recommending performance: a comparison

are sufficient for AEGCN to converge. It is worth to mention that, the best hyper-parameters for different missing rates need to be retuned.

5.3.3 Overall Recommending Performance Comparison (RQ1)

Figure 5.2 shows the recommending performance w.r.t ndcg of all models from 0% to 99% randomly missing in all attributes. We can see that AEGCN consistently outperforms all baselines on all three datasets for all missing rates, demonstrating its high effectiveness with simple designs. Table 5.3 shows the detailed results for specific cases when the attributes are complete and when 99% attributes are missing. Please note the performance of pure CF is not affected by missing rates. The detailed results for the other missing rates are omitted due to limited space.

Performance Comparison w.r.t. Missing Rate

Figure 5.2 shows that the performance of all attribute-enhanced algorithms generally decrease as the missing rate increases. However, AEGCN decreases much less than other attribute-enhanced methods. As the missing rate increases, the relative improvement compared with other algorithms becomes more obvious. We present the detailed results for 0% and 99% missing rates in Table 5.3.

- When the missing rate is 0%, the best baseline for Yelp-OH and Yelp-NC is DIN while for Tianchi is LightGCN. When missing rate is 99%, the best baseline for all datasets is LightGCN. NFM is the strongest attribute-enhanced baseline for Tianchi for all missing rates.

- When the missing rate is 0%, AEGCN improves over the strongest attribute-enhanced baselines w.r.t. ndcg@20 by 0.5%, 6.5%, and 59.2%, in Yelp-OH, Yelp-NC, and Tianchi respectively. When the missing rate is 99%, AEGCN achieves improvements over the strongest attribute-enhanced baselines w.r.t. ndcg@20 by 26.3%, 14.1% and 247.2% in Yelp-OH, Yelp-NC, and Tianchi respectively.

- When the missing rate is 0%, AEGCN outperforms the best pure CF baseline, LightGCN, by 6.4%, 14.5% and 11.0%. When the missing rate is 99%, LightGCN is the best baseline for all datasets. AEGCN still outperforms by 4.1%, 9.6% and 3.4% w.r.t. ndcg@20 on Yelp-OH, Yelp-NC and Tianchi respectively. AEGCN only decreases by 2.1%, 9.6% and 6.8% When the missing rate increase from 0% to 99% respectively. The results demonstrates that AEGCN is significantly better than the other methods, especially when the attributes are seriously missing. This is because: 1) AEGCN involves more attribute information to recommending task by profiling simultaneously; 2) Multi-task learning on the two related tasks – recommending and profiling – improves the performance.

Performance Comparison w.r.t. Interaction Sparsity Levels

One of the main motivations for introducing attributes into recommendation is to alleviate the Interaction Sparsity problem. Many users only have few interactions. It is hard to learn these users' preference over items. Here We investigate: 1) whether attributes can help to alleviate the interaction sparsity problem, and 2) how AEGCN performs with missing attributes. We conduct experiments over user groups of different sparsity levels. The test set is divided into three groups based on the interaction number per user. The sizes of interactions in all groups are almost equal. For example, the interaction numbers per user of each group are less than 3, 7 and 752 in Yelp-OH, respectively. Figure 5.3 shows the results w.r.t. ndcg@20 on different groups in all datasets when missing rate is 99%. A similar trend can be found in recall and omitted for space. From Figure 5.3, we can observe that:

- AEGCN consistently yields the best results on all user groups of all datasets, even there is only 1% attributes left. The best baseline on all datasets is a pure CF method, LightGCN, which is far better than the attribute-enhanced baselines. These results demonstrate that: 1) even a small portion of attributes can facilitate the preference learning; 2) Compared with other attribute-enhanced methods, AEGCN is more capable of alleviating the interaction sparsity problem when the missing rate is high.

- It is worthwhile to point out that almost all methods including AEGCN perform worse on the densest user group of Yelp-OH and Yelp-NC, compared with the first two sparser groups. There are two possible reasons for this. First, the most inactivate users of Yelp may prefer to visit a small number of top recommended local businesses. For example, the number of different items visited by the first group of Yelp-OH is only half of the third group. As a result it is easier to learn preference for the the first group. Second, the interactions of the densest user group actually are not so "dense". For example, as we checked, more than 50% of users in the "densest" group of Yelp-OH and Yelp-NC have less than 10 interactions. It is still hard for CF methods to learn these users' preference only through interactions. However, compared with the most inactive users, their preferences are not limited to some top recommended locations. Therefore, these users become the hardest group for CF methods.

5.3.4 Effects of Different Attributes (RQ2)

From Table 5.3 and Figure 5.2, we can find that the best baselines are not always attribute-enhanced methods, even when the missing rate is low. For instance, on Tianchi dataset, LightGCN and NGCF outperforms all the attribute-enhanced baselines even the missing rate is 0%. And on Tianchi dataset, AEGCN also achieves the largest relative improvements against other attribute-enhanced methods. We conjecture that it is caused

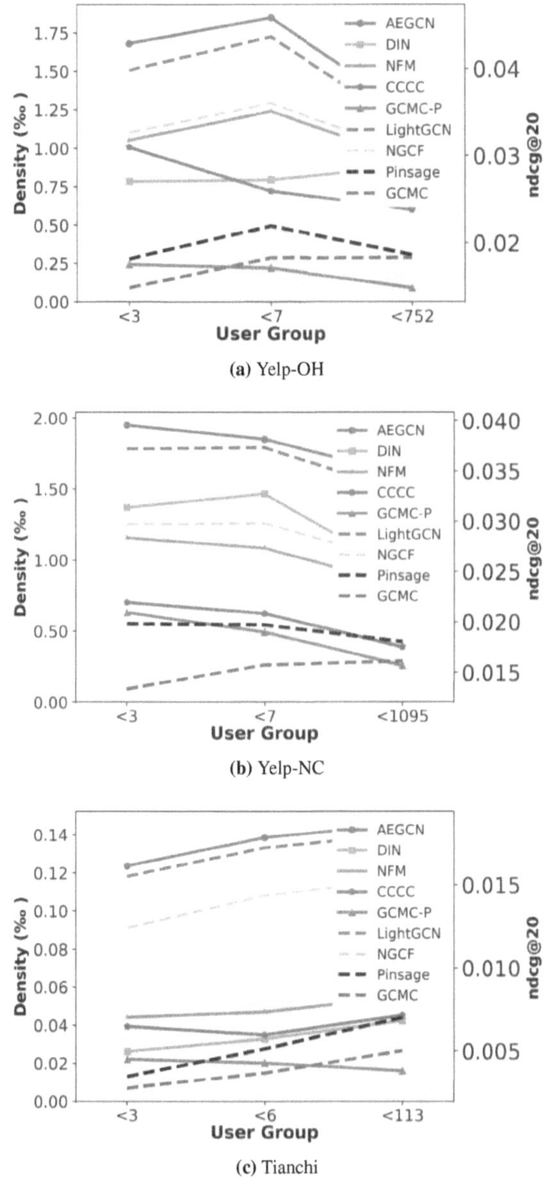

(a) Yelp-OH

(b) Yelp-NC

(c) Tianchi

Figure 5.3: Performance comparison over the sparsity distribution of user groups when the missing rate is **99%**. The background histograms indicate the density of each user group; meanwhile, the lines demonstrate the performance w.r.t. ndcg@20

by the fact that the attribute of Tianchi dataset is not so "effective", i.e., less relevant to the predicted items/user.

Strong and Weak Attributes. Our main target is to improve the performance of pure CF methods with attributes. Not all attributes lead to a better performance compared with pure CF methods. Some attributes are too weak that their contribution can be neglected. In this work, we use a simple method to quantify the effectiveness of different attributes in this work. We compare the performances of FM [75] with or without attributes. FM without attributes can be seen as a pure CF method [75]. If the performance of FM increases more than 10% after taking attributes as input, then the attributes are referred as strong attributes. By this standard, the attributes of Yelp-OH and Yelp-NC are all strong attributes while Tianchi is weak attributes.

Complete Strong Attributes. If there are strong and complete attributes, the attribute-enhanced algorithms generally perform better than pure CF methods. For example, DIN outperforms LightGCN by 6.4% and 7.5% w.r.t. ndcg@20 on Yelp-OH and Yelp-NC, respectively. And GCMC-P also outperforms GCMC by 39.52% and 31.75% w.r.t. ndcg@20 on Yelp-OH and Yelp-NC. These results demonstrate that strong attribute can effectively improve the performance of recommendation systems.

Incomplete Strong Attributes. If the missing rate of strong attributes increases, the performance of attribute-enhanced algorithms decreases significantly. It is reasonable because their performance heavily relies on the attributes and their interactions. For example, from 0% missing to 99%, DIN decreases about 25.8% and 10.4% w.r.t. ndcg@20 on Yelp-OH and Yelp-NC. The attribute-enhanced methods may be even weaker than pure CF methods if the missing rate of strong attribute become larger. This is because the complexity or noise caused by missing attributes becomes more important than the benefits of non-missing attributes.

Table 5.4: Best λ for Different Missing Rates

	Yelp-OH		Yelp-NC		Tianchi
	λ_u	λ_u	λ_u	λ_i	λ_i
0	0.001	0.1	0.01	0.001	0.01
0.1	0.1	0.01	0.1	0.1	0.01
0.2	0.001	0.01	0.01	0.001	0.01
0.3	0.01	0.001	0.1	0.001	0.001
0.4	0.001	0.001	0.001	0.001	0.001
0.5	0.1	0.001	0.001	0.01	0.001
0.6	0.01	0.01	0.01	0.001	0.001
0.7	0.01	0.01	0.001	0.001	0.001
0.8	0.001	0.01	0.01	0.001	0.001
0.9	0.01	0.001	0.01	0.001	0.001
0.99	0.001	0.001	0.001	0.001	0.001

Table 5.5: Profiling Performance (U – user profiling; I – item profiling)

	Yelp-OH(U)		Yelp-OH(I)		Yelp-NC(U)		Yelp-NC(I)		Tianchi(I)	
	F1	MCC	F1	MCC	F1	MCC	F1	MCC	F1	MCC
	0% missing attributes									
NFM	0.441	0.526	0.360	0.492	0.295	0.387	0.158	0.234	0.247	0.391
LightGCN	0.496	0.522	0.414	0.512	0.372	0.390	0.188	0.186	0.248	0.391
AEGCN	**0.535**	**0.555**	**0.431**	**0.515**	**0.418**	**0.377**	**0.200**	**0.234**	**0.306**	**0.432**
	99% missing attributes									
NFM	0.041	0.060	0.060	0.095	0.119	0.166	0.266	0.370	0.207	0.412
LightGCN	0.051	0.080	0.066	0.128	0.158	0.234	0.309	0.391	0.248	0.424
AEGCN	**0.059**	**0.073**	**0.067**	**0.131**	**0.174**	**0.280**	**0.321**	**0.385**	**0.267**	**0.469**

Weak Attributes. If the attributes are weak, LightGCN is the best baseline even there are no missing at all. There is only one item attribute in Tianchi, which is difficult for attribute-enhanced baselines to model effective feature interactions. Instead, the pure CF methods, which mainly focus on interactions, perform much better. Among pure CF methods, LightGCN outperforms NGCF, Pinsage and GCMC, which is consistent with the results of [45, 98]. AEGCN is based on LightGCN. So its relative improvements against other attribute-enhanced baselines are the largest in Tianchi dataset. AEGCN outperforms LightGCN due to two reasons: 1) AEGCN introduces attributes into LightGCN, getting better representation for inactive users; 2) multi-task learning itself is more effective for two related tasks.

5.3.5 Study of AEGCN (RQ3)

As the auxiliary profiling component plays a vital role in AEGCN, we investigate how it affects the recommending performance. We mainly focus on the influence of the most important hyper-parameters: loss weights λ.

Table 5.4 shows the best λ_u and λ_i for all datasets with different missing rates. λ significantly affects the performance of AEGCN. They need to be re-tuned on new datasets or new missing rates. Note that λ in Table 5.4 are only for the recommending tasks. If the profiling task is of higher priority, λ needs to searched in another range, which will be discussed in the next section. From Table 5.4, we can see that:

- The best λ for all datasets and missing rates are less than 1. This is because λ is actually a weight of profiling task, which controls the relative importance of the profiling task in the multi-task model. With a λ less than 1, the multi-task model will focus more on optimizing the loss of recommending task. In particular, the best λ_i for all missing rates are even smaller than 0.1. This mainly because that the loss of its profiling task is larger than recommending task by more than one order of magnitude. So it needs much smaller λ_i to control the weight of profiling task.

- The best λ become much smaller when the missing rate is higher: 1) the best λ of 99% missing rate for all datasets is 0.001; 2) all the best λ of more than 60% missing rate for Yelp-OH and Yelp-NC is between 0.01 and 0.001; 3) the best λ of Tianchi is 0.001 after missing rate just reaches up to 30%. This can be attributed to the following reasons. First, as the missing rate increases, the useful attribute information from the auxiliary profiling tasks generally becomes less, which makes the profiling components less important to recommending task. Hence, a larger λ is not necessary. Second, a higher missing rate also makes it harder to estimate attributes, which leads to higher errors and losses in profiling tasks. As a result, the model needs a much smaller λ to reduce the negative impact of profiling tasks for recommending tasks.

5.3.6 Profiling Performance (RQ4)

λ **needs to be retuned for Profiling**. In this section, we compare the performance of AEGCN with other single-tasks on profiling tasks. The parameters discussed in previous sections are not suitable for profiling tasks. We find that AEGCN may perform worse than baselines with those parameters when the missing rate is high. This is because that, the multi-task model does not pay enough attention to the profiling part, when focusing on recommending tasks. As shown in Table 3, all the best λ is less than 1, and only 0.001 if the missing rate is 99%. So the parameters need to retuned for profiling tasks. Especially, λ is searched in {0.1, 1, 10, 100}.

Experiments Settings. The profiling dataset are split into training, validation and testing sets by 75%, 5% and 20%. And we mainly compare the performance with two best baselines: NFM and LightGCN. The final user embedding and item embedding of these methods are fed into a Dense layer and softmax layer for profiling, which is the same as AEGCN. The distribution of attributes for all datasets is highly uneven. Take the user attribute of Yelp-NC as an example, there are 1309 person in the largest category and only 1 person in the smallest category. And there are 18 categories which have less than 50 persons. To evaluate the effectiveness of profiling on imbalanced datasets, we adopt two widely-used evaluation metrics: F1 [60] and Matthews Correlation Coefficient (MCC) [15]. Generally, the profiling performance is better if values of these two metrics are larger.

Results. Table 5.5 shows the detailed metrics of profiling tasks when the missing rates are 0% and 99%, respectively. The performance on other different missing rates follows the similar trend and are omitted for

 Chapter 5 Multi-Task-level Problem: Improving User-Attribute-Enhanced tasks by

space. The results also show that AEGCN outperforms LightGCN and NFM, demonstrating that the multi-task learning method can also help on profiling tasks. It is worth to mention that the best hyper-parameter λ for profiling are all over 10. And most of the best λ are very large (100) when the missing rate is 99%. Besides, the category numbers for different missing rates are different. There are less categories to predict when the missing rate is very high. So the profiling performance does not simply decrease when the missing rate increases. For example, in the Tianchi dataset, there are 301 and 130 item categories when missing rate are 0% and 99%, respectively. We find that the most difficult missing rates for profiling tasks are between 50% and 80%. In this range, the category numbers are still close to those when attributes are complete, while a large part of attributes are already missing. So there are only a small number of instances in many categories. This is hard to estimate the attributes for all methods. This problem will be explored in our future work.

5.4 Chapter Summary

In this chapter, we propose AEGCN, a multi-task attribute-enhanced GCN-based CF method, which improves the performance of recommending task by simultaneously estimating missing user/item attributes. The experiments show that AEGCN consistently performs better than state-of-the-art CF methods. Especially, when a large part of user/item attributes are missing, the relative-improvement compared with attribute-enhanced methods significantly increases. To the best of our knowledge, this work is a first attempt to investigate how to exploit incomplete attributes in GCN-based CF methods with the help of UAI.

Chapter 6

Chapter 6

Conclusion

In this chapter, we first summarize the three works on the user attribute inference via mining user-generated data in this book. Then we will discuss the future work.

6.1 Summary

In this book, we study three open problems on the user attribute inference via mining user-generated data:

- For the single-attribute-level problem, we aim to introduce human mobility data into SES inference. Previous SES inference works are based on users' social media data and overlooked the people's mobility data. The Shanghai subway smart card data is chosen as a case study. Through mobility pattern analysis, housing price data mining and income-housing price survey, we construct the SES label for more than 700 thousand users using the house-price level of their estimated home location. Then we design a new sequential functional mobility feature that consider people's dynamic mobility pattern between different city function areas. A deep learning model, S2S (short for SCD to SES), is proposed to estimate their SES-level by combining both the traditional statistical mobility feature and sequential functional mobility feature. In the end, experiments show that mobility data can be used to estimate the SES level and much better than random guess). Besides, the sequential features are clearly better than traditional general features.

- For the multi-attribute-level problem, we want to improve the accuracy of multiple SEA inferences from limited data sources like the home location. Previous SEA inference works are based on users' tweets content or mobile phone usage data. To build a dataset consisting of SEA and home location, we collected people's socioeconomic attributes and their home locations in 9 provinces and 85 cities of China. To get more information from home location, various kinds of SEA-related home-based features are designed, like housing prices, county-level income, and urban types. Corresponding data are mined from various websites including government statistic websites and commercial housing websites. Then an FM-based multi-task learning method named H2SEA is proposed to model both second-level feature interactions to further improve the prediction accuracy. To test the performance of the proposed model and feature, extensive experiments are conducted on the collected datasets. The results show that the home-based features and proposed method can clearly improve the performance in predicting people's SEAs, which outperforms the compared methods in terms of various metrics like AUC and F1.

- We expand the focus from UAI to the cooperation of UAI and UAE. Our purpose is to improve the performance of CF recommender system with help of UAI. To lower the cost, CF recommendation methods usually neglect UAI and simply use unknown tags as substitutes for missing attributes. We first quantify the negative impact of ignoring UAI based on 3 real-world datasets. The performances are decreased by more than 10% in all datasets when the missing rate is more than 90%. Then we propose AEGCN, a multi-task attribute-enhanced GCN-based CF method. It improves the performance of recommending tasks by simultaneously estimating missing user/item attributes. The experiments show that AEGCN consistently performs better than state-of-the-art CF methods. Especially, when a large part of user/item attributes is missing, the relative-improvement compared with attribute-enhanced methods significantly increases.

From the first to the third work, we try to expand UAI: 1) from one-attribute-prediction to multi-attribute-prediction and finally multi-task framework; 2) from only serving UAB to serve both UAI and UAE tasks. The proposed methods and corresponding discussions are not just applicable to each case study. They also belong to a general aim and on-going effort of UAI community: provide a general UAI framework. The general framework not only covers many kinds of user-generated data sources and user attributes but also can serve various kinds of UAB and UAE tasks simultaneously.

6.2 Future Work

As mentioned above, this book mainly tries to expand the serving targets of UAI from a single-attribute to multi-attributes inference, and even the other kinds of tasks like the recommender system. There are two general future directions we can consider. First, we still need to continue to expand the serving targets. Second, we should consider expanding kinds of input data sources for UAI in the future works. Next, we discuss some specific extension directions for each work.

- For the first work, the ground truth dataset is the first problem. Because we use the house price of people's living area as the ground truth. We cannot leverage some important features (e.g., favorite locations and housing price level of their working area) in estimating SES. We plan to conduct a detailed SES survey of a reasonable scale to build a more precise model between SES and mobility as future work. The second direction is to combine mobility, cellphone records, or even new kinds of data sources into

SES prediction. The third direction is to explore whether more advanced models proposed in recent years like the graph-based sequential model can be used to further increase the accuracy.

- For the main problem of SEA prediction, the first expanding direction is also the dataset. Collecting ground truth and building basic feature datasets cross China cost us a lot of time. We are not able to collect data in other countries. As a result, some of the conclusions may not hold in the other areas. For example, housing prices are not so effective in our experiments, but this may be different in other countries. What's more, the complexity of the model is limited by the scale of datasets. We tried to leverage a more sophisticated model however suffered serious overfitting problems caused by the datasets. In the future, we plan to collect more datasets and develop a more general model which can be applied in different countries.

- For the issue of the third work, there are also some potential new issues that need to be addressed. First, compared with attribute-enhanced methods like DIN and NFM, the relative improvements of AEGCN is not very obvious for strong and low-missing attributes. As the next step, we will improve AEGCN by exploiting complex feature interactions. Second, though the improvement is clear, the explanation of how multi-task learning affects the results is still not clear. We are not sure the exact contribution of user ID, original attributes or estimated attributes in the final results. We need to leverage new methods to distinguish their contribution. Third, we plan to consider cold-start user problem in future work, which means users without any interactions at all. Finally, we want to try to expand the framework to other UAE tasks like the precise advertisement.